WATER RESOURCE PLANNING, DEVELOPMENT AND MANAGEMENT

DRINKING WATER AND WATER MANAGEMENT

NEW RESEARCH

Water Resource Planning, Development and Management

Additional books in this series can be found on Nova's website under the Series tab.

Additional e-books in this series can be found on Nova's website under the e-book tab.

WATER RESOURCE PLANNING, DEVELOPMENT AND MANAGEMENT

DRINKING WATER AND WATER MANAGEMENT

NEW RESEARCH

MASON GIANNINO
EDITOR

New York

For permission to use material from this book please contact us:
Telephone 631-231-7269; Fax 631-231-8175
Web Site: http://www.novapublishers.com

Additional color graphics may be available in the e-book version of this book.

Library of Congress Cataloging-in-Publication Data

ISBN: 978-1-63117-523-7

Published by Nova Science Publishers, Inc. † New York

CONTENTS

PREFACE

Water is an essential natural resource for life. Nowadays, water contamination is a common issue due to a variety of sources but mainly of anthropogenic origin, such as urban and industrial wastes. Methods for evaluating water availability and reliability in supplying needs for agricultural, municipal, and industrial water use, environmental flows, electric energy generation, and reservoir storage are described in this book. This book also discusses chemical threats in drinking water; removal of radioactivity from drinking water; and irrigation agriculture.

Chapter 1 - This chapter outlines capabilities for assessing river/reservoir system water availability and supply reliability, focusing on methods incorporated in the Water Rights Analysis Package (WRAP) modeling system, which is generalized for application anywhere in the world. The generalized modeling system contributes greatly to water management in Texas in the United States and continues to be expanded to address additional water management issues. The Texas Water Availability Modeling (WAM) system consists of WRAP and WRAP input datasets for all of the river basins of the state. The WAM system supports a regional and statewide planning process, administration of water allocation systems, and other water resources management functions. Activities of numerous water management entities operating 3,400 reservoirs and other facilities in accordance with a USA-Mexico treaty, five interstate compacts, two water right permit systems with over 6,000 active permits, and other institutional arrangements are simulated. A comprehensive array of water supply reliability, hydroelectric energy reliability, reservoir storage frequency, and stream flow frequency metrics have been incorporated in WRAP and applied in the Texas WAM System. The reliability and frequency analysis methods adopted in WRAP are applicable to other modeling systems as well as other regions of the world.

Chapter 2 - Vulnerability of drinking water distribution systems (DWDS) is one of the main issues of concern to regulatory agencies and water utilities. Indeed, the threat of a deliberate (or even accidental) contamination of water supplies is real and is permanently under consideration. Although measures against such contamination events have been intensified in the last few years, major knowledge gaps still exist in what concerns the feasibility of the detection and remediation methods, especially those related with contamination by chemicals. This has been recently addressed in a European Project named SecurEau (www.secureau.eu) whose main objective was to launch an appropriate response for rapidly restoring the use of a DWDS after a contamination event. This chapter intends to present the developments made concerning the remediation methods, proposing several decontamination procedures, based on advanced oxidation processes (AOPs), for the treatment of waters/solids contaminated with key chemicals. Based on several criteria, three model compounds were selected: an organophosphate (chlorfenvinphos - CFVP), an herbicide (paraquat - PQ) and a brominated flame retardant (2,2´,4,4´,6-pentabromodiphenyl ether (BDE-100)).

The treatment of waters contaminated with PQ was studied by both Fenton and photo-Fenton process under off-line conditions (batch reactor) and by Fenton´s process under on-line conditions (i.e., in-situ of a pilot loop reactor, simulating a DWDS). Out of the pipelines, the photo-Fenton approach has great advantages over Fenton´s process since it allows a significant increase on the PQ degradation rate and, particularly, on the mineralization degree (96% after 1 h of reaction). Concerning the in-situ treatment, the Fenton's reagent proved to be effective for the treatment of PQ-contaminated waters, and it was shown that some deposits and pipe materials could be used as catalysts, avoiding the addition of other chemicals inside the network.

Water and deposits contaminated with CFVP can also be treated in-situ by Fenton's reagent making use of the active iron available either on pipe materials, on deposits attached to them or on loose deposits as catalysts of the process.

The direct photolysis methodology was successfully applied to the off-line remediation of BDE-100 contaminated waters. Batch experiments with pure water revealed a high BDE-100 degradation profile for the first five minutes of exposition to UV-light (1.6×10^{-6} Einstein/s).

Therefore, it has been proved that several solutions exist using AOPs for the effective treatment of waters and deposits contaminated with various chemicals, with different properties; this might be an important step forward towards chemical threats in drinking water distribution systems.

Chapter 3 - Water is an essential natural resource for life. Nowadays, water contamination is a common issue due to a variety of sources but mainly of anthropogenic origin, such as urban and industrial wastes.

Chlorination is probably the most common process used worldwide to treat waters intended to human consumption, because it efficiently removes pathogenic microorganisms like those responsible of malaria or typhus. However, the chlorine used in the process can react with the natural organic matter already present in waters, mainly humic and fulvic acids, originating disinfection byproducts (DBPs). The extent of such formation is also dictated by the bromine content of waters as well as by the acidity. Among disinfection byproducts, haloacetic acids (HAAs) and trihalomethanes (THMs) can be cited.

HAAs are highly polar compounds. Their structure is based in that of acetic acid molecule, but having the alpha carbon adjacent to the carbonyl with one, two or three halogens atoms (Br or Cl). There is a growing concern about the presence of HAAs in waters. Current in vivo studies with mice have proved toxic and carcinogen effect. Indeed, the International Agency of Research on Cancer (IARC) has classified four HAAs as possible human carcinogenic, and the United States Environmental Protection Agency (US-EPA) classifies one HAA (trichloroacetic acid) as possible human carcinogen and another HAA (dichloroacetic acid) as probable human carcinogen.

In this sense, there is a need of analytical methods for the efficient determination of HAAs in water. Current analytical methods take use of gas-chromatography (GC), high-performance liquid chromatography (HPLC), and ion-chromatography (IC), all in combination with a variety of detection systems. When GC is utilized, a derivatization step is needed to increase the volatility of HAAs (forming esters in the majority of cases). Afterwards, the extraction is normally conducted using headspace (HS) or solid-phase microextraction (SPME). When HPLC is used, the most common strategy is to employ hydrophilic interaction liquid chromatography (HILIC), generally in combination with a supported liquid membrane microextraction (SLMME) method. When IC is used (with or without a suppressor column), the extraction strategy normally involves solid-phase extraction (SPE). This book chapter intends to give an overview of the most recent trends in analytical methods to monitor HAAs in waters intended for human consumption, after their chlorination.

Chapter 4 - Water used for human consumption can present different types of contaminants, mainly chemical and biological. The routine procedures of Drinking Water Treatment Plants (DWTP) remove the majority of these contaminants. However, the radioactivity content present in the water remains mostly unaffected by these routine procedures. The sources of radionuclides can be either anthropogenic or naturally occurring, the latter being closely related to the geological characteristics. The naturally occurring radionuclide content of surface and ground waters are the result of the incorporation into the original rainwater of diverse chemical elements found in the soils through which the water flows. In the case of areas not affected by serious radioactive contamination events, the natural source is the most important. Therefore, the drinking water intake can pose a radiological hazard, and it is regulated by establishing a series of quantitative thresholds regarding the Total Indicative Dose (TID), and other parameters such as gross alpha and beta content. There are a lot of radiochemical procedures that are able to eliminate practically the main radioactive ions solved in water. Their generalized application is, however, limited by their high selectivity for each element or group of elements and their relatively high costs. It occurs often that the water treated acquires chemical characteristics which make it not drinkable. The traditional procedures of DWTPs can be also modified to maximize the removal of radionuclides, mainly uranium and radium, while the output water remains within drinkable limit, also established by legislation. These modifications are obtained by analyzing the best conditions for all steps in the process, from filtration to coagulation-floculation, for different types of reagents at varying concentrations and pH. Then, they were tested successfully in operating DWTPs, which provided the population with drinkable water which radionuclide content was much lower.

Chapter 5 - Endocrine disrupting compounds represent a wide variety of pollutants, natural and synthetic, among which the steroidal hormones exhibit the highest estrogenic activities.

Since water monitoring for these compounds can be challenging, sensitive analytical techniques are required but also the sampling technique has to be considered. Most conventional screening methods in water involve spot sampling, providing only a snapshot of the levels of pollutants at the time of sampling. However the concentrations of pollutants can fluctuate very rapidly depending on occasional releases or environmental conditions and frequent sampling is often required. Passive sampling represent an alternative approach: it provides the average concentration of the monitored compounds in the site as a function of the exposure time, enabling also the in-situ preconcentration of target analytes.

In the author's laboratories a fast analytical method using LC-MS/MS was developed to determine five EDC's, often included in studies regarding EDC monitoring in waters, since they are considered representative of phenolic compounds and steroid estrogens: 4-n-nonylphenol, bisphenol A, estrone, 17β-estradiol and 17α-ethinylestradiol. A case study on raw surface water and drinking water was taken into account using the approach of water passive sampling: obtained results are here presented and discussed.

Chapter 6 - The environmental persistence and fate of quinoline in coking wastewater is of particular interest because it mainly contributed to COD concentration in effluent, and made the effluent discharge not be well satisfied with sewage discharge standard (GB 13456-2012). In order to efficiently remove quinoline from coking wastewater, the authors employed biological strengthening technology to degrade quinoline, investigated the influence of environmental factors on the degradation performance of quinoline and its application in treating real coking wastewater. A highly efficient quinoline-degrading bacteria strain *Comamonas testosterone sp.* Y-01 was screened and isolated from activated sludge taken from coking wastewater. Biodegradation experiments showed Y-01 had strong ability for quinoline degradation, degrading as much as 100 $mg{\cdot}L^{-1}$ of quinoline within 12 hours and the biodegradation efficiency of quinoline reached 96-99%. Y-01 can almost completely remove quinoline in an environment with pH values between 4 and 10 at 30 °C. External inorganic-nitrogen and carbon sources apparently had no influence on the quinoline degradation ability of this strain. External phenol, however, can accelerate the biodegradation of quinoline. When this strain was mixed with phenol-degrading bacteria, Y-01's efficiency in degrading quinoline significantly increased. The predominant bacterium Y-01 was employed to replace the traditional complex microorganism on the carriers in moving-bed biofilm reactor (MBBR) to treat real coking wastewater. 74.49% quinoline is removed within 46 h, increased by 28.59-55.49% comparing to MBBR with traditional complex microorganism. Adding phenol degrading bacterium into the new dominant bacterium system in MBBR could weaken the inhibition on quinoline degrading bacterium in coking wastewater treatment, leading to a rise of 19.07% compared to without adding phenol degrading bacterium system, and 47.66-74.56% to MBBR with traditional complex microorganism. The above results will help us understand the biodegradation process and facilitate the further improvement in the practical applications of biological strengthening technology.

Chapter 7 - Sustainable irrigation agriculture is required to provide much-needed food and fibre but is threatened by reduced water supply and increasing competition for environmental needs and urban and industrial water demand. Policy on water allocation often pits the needs of irrigation against competitive demands. This chapter explores innovative options for avoiding or mitigating this conflict. Examples include: the strategic location of forestry to minimise interception of water in the landscape; investment in infrastructure to provide environmental benefits with minimal use of more water; reducing water loss by optimising river operations; use of water sharing agreements to provide greater certainty to irrigators and investors; market-based trading arrangements to provide flexibility to irrigators to deal with water scarcity and climate variability; improved water use efficiency both on farm and in water supply and delivery systems; and more efficient water use and re-use in cities and industry. The water cycle and water auditing provide the biophysical context for developing options by governments, policy makers and communities as they adapt to change. Processes of public participation remain a challenge but good progress is being made.

Chapter 8 - The idea of resilience captures individual, sectoral or community capacity to tackle disruptive events that change their status quo, and to successfully transform themselves into new states of existence. During the Millennium drought in Australia, many irrigators faced severe water shortages and unusual government market activity which constituted significant disruptive events. Coincidentally, water reform prior to these events provided scope for water trading that, together with other adjustment packages and support, arguably offered a major method for achieving resilience among affected irrigator groups. This chapter reviews those events and the resilience outcomes from water trade activity in Australia's Murray-Darling Basin during the Millennium drought, with an objective of attempting to tangibly link irrigator resilience to water trade availability in the Basin. A methodological approach for testing irrigator resilience outcomes is proposed and specified for subsequent modelling using irrigator water trade data collected during the Millennium drought. Knowledge of the properties that confer resilience in irrigation systems may enable other resource managers, users and communities to design and implement appropriate polices that minimize social impacts while maximizing sustainable water management and ecosystem goods and services outcomes.

Chapter 9 - The continuous over-exploitation of groundwater resources due to the increased demand for irrigation water has lowered water tables in many riparian areas, threatening thus many water-stressed regions. Simulation models are useful tools for efficient water management of those areas, especially when integrating economic and hydrological information. This paper focuses on the simulation of groundwater use on a basin-wide scale, in the Moudania basin, in Greece. Namely, a two-dimensional transient flow simulation in the aquifer of Moudania was performed, aiming first of all to assess the impact of hydraulic head distribution on the pumping depth of groundwater, as well as, its relative (long-term) effect on irrigation pumping costs. In this context, a steady-state simulation was first performed in order to get the values for certain parameters (e.g., hydraulic conductivity), as well as the initial conditions for the transient simulation. These two simulations, referring as the basic simulation, were carried out using the modular three-dimensional finite-difference groundwater flow model MODFLOW.

The economic cost of groundwater abstraction was then estimated based on the cost of energy (electricity) for pumping irrigation water. The economic analysis results showed an association of pumping costs to the depth of water abstraction, as well as to the current water consumption levels. Moreover, the range of future marginal pumping costs seems to vary spatially within the basin area, as it is directly linked to the drawdown levels. However, according to the author's estimates, pumping costs constitute and will continue to constitute only a small part of farmers' production costs, leading thus to inefficient use of groundwater resources and thus to depletion.

The simulation model was also used to evaluate an alternative policy scenario, with respect to the hydraulic head of the aquifer and the agricultural production of the study area (i.e., farmers' income). This alternative scenario refers to the formulation and solution of the inverse problem, by assigning hydraulic head values that are higher than a predetermined level and thereafter by determining the pumping rates of irrigation wells. The final outcome of this process would be a spatially diverse system of water permits that would minimize the resource cost, which may arise due to the depletion of the aquifer.

In: Drinking Water and Water Management ISBN: 978-1-63117-523-7
Editor: Mason Giannino

Chapter 1

METRICS OF RIVER/RESERVOIR SYSTEM WATER AVAILABILITY AND SUPPLY RELIABILITY

Ralph A. Wurbs
Zachry Department of Civil Engineering,
Texas A&M University, US

ABSTRACT

This chapter outlines capabilities for assessing river/reservoir system water availability and supply reliability, focusing on methods incorporated in the Water Rights Analysis Package (WRAP) modeling system, which is generalized for application anywhere in the world. The generalized modeling system contributes greatly to water management in Texas in the United States and continues to be expanded to address additional water management issues. The Texas Water Availability Modeling (WAM) system consists of WRAP and WRAP input datasets for all of the river basins of the state. The WAM system supports a regional and statewide planning process, administration of water allocation systems, and other water resources management functions. Activities of numerous water management entities operating 3,400 reservoirs and other facilities in accordance with a USA-Mexico treaty, five interstate compacts, two water right permit systems with over 6,000 active permits, and other institutional arrangements are simulated. A comprehensive array of water supply reliability, hydroelectric energy reliability, reservoir storage frequency, and stream flow frequency metrics have been incorporated in WRAP and applied in the Texas WAM System. The reliability and frequency analysis methods adopted in WRAP are applicable to other modeling systems as well as other regions of the world.

1. INTRODUCTION

Methods for evaluating water availability and reliability in supplying needs for agricultural, municipal, and industrial water use, environmental flows, electric energy generation, and reservoir storage are described in this chapter. Effective management of the water resources of a river system requires an understanding of the amount of water available at various levels of reliability under various conditions. Water availability depends upon river basin hydrology, constructed facilities, institutional water allocation systems, water management practices, and basin-wide water demands. Stream flow and other hydrologic variables are highly stochastic characterized by great variability including the extremes of severe droughts and floods. The future is always uncertain. Thus, the availability of stream flow and reservoir storage and associated water supply capabilities must be expressed in terms of probability, frequency, percent of time, risk of shortage, and/or supply reliability.

The general modeling and analysis strategy outlined in this chapter consists of simulating a specified scenario of water resources development, management, and use during a postulated repetition of natural river basin hydrology. Supply reliability and storage and flow frequency metrics are developed from the results of the simulation model. The river/reservoir/use system being simulated may range in complexity from a single water user supplied by a single reservoir to complex systems of numerous multiple-purpose reservoirs and water users. Long-term reliability and frequency analyses may be performed in planning studies and administration of water allocation systems. Short-term reliability and frequency analyses support short-term drought management decision processes and operational planning studies.

The modeling and analysis methods outlined in this chapter are implemented in the Water Rights Analysis Package (WRAP), which is a generalized modeling system for simulating water resources development, management, allocation, and use in river basins located anywhere in the world. WRAP is designed for assessing reliabilities in meeting water supply, hydroelectric power, and environmental flow needs and also includes optional capabilities for tracking salinity and simulating reservoir operations for flood control. The WRAP software and user documentation are available free-of-charge at: http://ceprofs.tamu.edu/rwurbs/wrap. htm.

WRAP is routinely applied in the Texas Water Availability Modeling (WAM) system in regional and statewide planning studies, administration of the water rights permit system, and other water management activities. The Texas WAM system created and maintained by the Texas Commission on Water Quality (TCEQ) and its partner agencies and contractors consists of the generalized WRAP developed at Texas A&M University and WRAP input datasets for the 23 river basins of Texas. WAM information including WRAP input datasets for the river basins of Texas is available at: http://www.tceq.texas.gov/permitting/water_ rights/wam.html.

The literature on modeling and analysis of river/reservoir systems is massive, including thousands of journal and conference papers and technical reports. Pioneering efforts in computer simulation of reservoir systems include U.S. Army Corps of Engineers studies of six reservoirs on the Missouri River initiated in 1953, International Boundary and Water Commission simulations of the Rio Grande in 1954, and a simulation study of the Nile River Basin in 1955 [1]. Several books on modeling and analysis of reservoir operations are available [2, 3, 4, 5]. Water resources systems books [6, 7, 8, 9] cover simulation, optimization, and stochastic analysis concepts relevant to river/river system reliability analyses. Labadie [10] summarizes the extensive and complex research literature on reservoir system optimization models. Wurbs [11, 12] presents a state-of-the-art reviews of reservoir system analysis from a practical applications perspective. Wurbs [12] compares WRAP and other generalized reservoir/river system simulation models.

2. Water Rights Analysis Package (WRAP) Modeling System

Development of WRAP at Texas A&M University began in the late 1980's sponsored by a cooperative research program of the U.S. Department of the Interior and Texas Water Resources Institute (TWRI). WRAP has been greatly expanded since 1997 under the auspices of the Texas Commission on Environmental Quality (TCEQ) in conjunction with implementing a statewide Water Availability Modeling (WAM) System. The Texas Water Development

Board, U.S. Army Corps of Engineers, and other agencies have also sponsored improvements to WRAP. The modeling system is documented by a set of TWRI technical reports [13, 14, 15, 16, 17, 18, 19] available at the TWRI website (http://twri.tamu.edu/ publications/reports/) as well as the author's WRAP website (http://ceprofs.tamu.edu/ rwurbs/wrap.htm).

WRAP simulates capabilities of river/reservoir systems in meeting specified water management, regulation, and use requirements for given sequences of naturalized stream flows and reservoir net evaporation less precipitation rates. Water management is combined with natural hydrology. Water managers are concerned with future not past hydrologic conditions. However, since the future is unknown, historical hydrology is used to capture the hydrologic characteristics of a river basin. The water management and use scenario might be actual current water use, projected future conditions, the premise that all permit holders use their full authorized amounts, or some other scenario of interest. Simulation results are organized in optional formats including tabulations and plots of entire time sequences, summary tables, water budgets, frequency relationships, and various types of reliability indices.

2.1. WRAP Software

The WRAP modeling system includes a set of executable Fortran programs that perform the functions outlined in Table 1. WinWRAP is a user interface for executing the programs on microcomputers within Microsoft Windows®. WinWRAP provides model-users a convenient environment in which to manage data files, WRAP programs, and auxiliary programs. WinWRAP connects WRAP executable programs with each other and with pertinent data files. Simulation results are manipulated or plotted with Microsoft Excel® or HEC-DSSVue [20]. WRAP Display [21] is an ArcGIS® tool for spatially displaying simulation results.

Program SIM performs the river/reservoir/use system water allocation simulation using a monthly time step. SIMD (D for daily) is an expanded version of SIM with daily time step, flow forecasting, routing, environmental pulse flow tracking, and flood control simulation features.

Program HYD is a set of computational routines for converting sequences of monthly gauged stream flows to naturalized flows and compiling sets of monthly net reservoir evaporation less precipitation depths. HYD output consists of hydrology input datasets for SIM. Program DAY has routines for synthesizing daily or other sub-monthly time interval flows from monthly naturalized flows and calibrating routing parameters. DAY output is input for SIMD.

Table 1. Water Rights Analysis Package (WRAP) Modeling System

1. WinWRAP is a user interface for managing programs and data files within Microsoft Windows.
2. Development of Hydrology Input Data for the Simulation Model
   - Program HYD described by the Hydrology Manual [16] develops and updates SIM input files containing monthly naturalized stream flows and reservoir evaporation-precipitation rates.
   - Program DAY documented by the Daily Manual [18] is used to calibrate routing parameters and otherwise compile daily hydrology input data for SIMD.
3. Simulation of the River/Reservoir/Water Use System
   - Program SIM performs monthly simulations [13, 14, 15].
   - Program SIMD performs daily simulations as described in the Daily Manual [18].
4. Tracking Salinity through the River/Reservoir System
   - Program SALT performs a salinity simulation as described in the Salinity Manual [17].
5. Post-Simulation Analyses of Simulation Results
   - Program TABLES reads SIM, SIMD, and SALT simulation results, performs frequency and reliability analyses, and creates a variety of tables to organize, summarize, and display simulation results [13, 14, 15, 17, 18].
   - HEC-DSSVue [20] reads SIM, SIMD, and SALT simulation results DSS files and DSS files created by TABLES and prepares plots and can also perform various statistical analysis.
   - The WRAP Display Tool [21] functions as a toolbar within the ArcMap component of ArcGIS® to spatially display SIM and SIMD simulation results.

Program SALT reads a SIM output file and salinity input file and tracks salt loads and concentrations through a river/reservoir system. SALT is a separate program that uses SIM simulation results in modeling salinity but does not affect the SIM simulation.

Program TABLES organizes the SIM, SIMD, and SALT simulation results and develops frequency relationships, reliability indices, and summary statistics. TABLES organizes simulation results into a variety of user defined tables and also creates data files for easy export to Microsoft Excel® or HEC-DSSVue.

2.2. Simulation Input Data

The spatial configuration of a river system is defined by a set of control points, with the next downstream control point being specified for each control point. All reservoirs, diversions, return flows, hydropower plants, environmental instream flow requirements, and other system components are assigned control point locations. Essentially any configuration of stream tributaries and conveyance systems may be modeled. There are no limits on the number of control points, reservoirs, or water rights that may be included in a dataset.

Hydrology input for the WRAP simulation model SIM consists of sequences of naturalized stream flows at primary control points and net evaporation less precipitation rates for all reservoirs. The hydrologic period-of-analysis should be sufficiently long, reflecting a full range of fluctuating wet and dry periods, to allow simulation results to be used to develop frequency relationships, reliability indices, and other statistics that characterize the water resources.

Observed historical stream flows are adjusted to develop a homogeneous set of flows representing a specified condition of river basin development. The extent of the adjustments varies depending on circumstances. For relatively undeveloped watersheds, little or no adjustments may be necessary. In extensively developed river basins, adjusting for the effects of all human activities is not feasible. Naturalized flows are typically developed by adjusting recorded flows at gauging stations to remove the impacts of major upstream reservoirs, diversions, return flows from surface and ground water sources, and possibly other factors. Other changes may also be considered. A watershed precipitation-runoff model may be applied to develop adjustments for changes in watershed land use and vegetative cover.

Primary control points are locations, usually gauging stations, for which naturalized flows are provided in a SIM input dataset. Naturalized flows at secondary control points are computed during a simulation. The model includes several alternative methods for transferring naturalized flows from gauged to ungauged sites. Flows may be distributed in proportion to drainage area with or without considering channel losses. SIM also includes an option based on the Natural Resource Conservation Service relationship between precipitation and runoff [16].

Channel losses may be considered in distributing naturalized flows from primary to secondary control points. Channel losses are also included in modeling the effects of diversions, return flows, and reservoirs on stream flows at downstream locations.

Water management, regulation, and use requirements, policies, practices, and facilities are described in terms of water rights. WRAP provides flexibility for modeling complex system configurations and operations. Extensive improvements to the generalized model were made in response to various situations encountered as each river basin was modeled for the Texas WAM System, providing flexibility needed to address diverse water management practices. Features for defining water use requirements and management practices in a simulation input dataset include:

- identifiers for aggregating simulation results for groups of related water rights
- locations of system components by control point
- priority specifications
- water supply diversion, environmental instream flow, and hydroelectric energy targets for each of the 12 months of the year and specifications for varying the water use targets as a function of reservoir storage contents or stream flow
- seasonal or annual limits on diversions, reservoir releases, or flow depletions
- return flow specifications in various optional formats
- conveyance of flow through pipelines and canals
- operating rules including multiple-reservoir system operations, multiple-purpose operations, multiple-owner reservoirs, off-channel storage, and constraints on depleting stream flows
- reservoir storage volume versus surface area and elevation relationships

2.3. Forms of Stream Flow

The WRAP modeling process consists of a series of adjustments to stream flow sequences covering the hydrologic period-of-analysis (simulation period). The Texas WAM System described later reflects simulation periods of 50 or more years for the various river basins and a monthly time step.

A simulation performed with SIM or SIMD results in three forms of stream flows at each control point: naturalized, regulated, and unappropriated. HYD provides capabilities for developing naturalized flows by adjusting sequences of observed monthly flows from a gauging station to remove the historical effects of water resources development and management. HYD, SIM, and SIMD include methods for transferring flows from gauged to ungauged control points.

A SIM or SIMD simulation begins with naturalized flows. In general, the terms *naturalized*, *unregulated*, or *unimpaired* refer to sequences of past stream flows adjusted to represent a specified condition of river basin development that includes either no human impact or some defined level of development. For the Texas WAM System, naturalized flows are ideally flows that would have occurred historically, in the absence of the water management activities reflected in the water rights input data, but with all other aspects of the river basin reflecting constant present conditions.

Regulated and unappropriated flows computed by SIM or SIMD reflect adjustments to naturalized flows for water right requirements representing a specified scenario of water resources development and use. Regulated flows are physical flows considering all water rights in the input dataset. Unappropriated flows are available for further appropriation after all the water rights receive their allocated share. Regulated flows may be greater than unappropriated flows due to instream flow requirements at the site or commitments to other rights at downstream control points.

The adjustments that convert naturalized flows to regulated flows include both stream flow depletions and return flows. Stream flow depletions are the quantities of water appropriated to meet water supply diversion requirements and refill reservoir storage. Return flows are added back to stream flows. Channel losses are considered as SIM and SIMD stream flow adjustments are cascaded downstream. Routing is performed for flow adjustments in SIMD.

2.4. Simulation Computations

SIM simulation computations are performed in a water rights priority loop that is embedded within a monthly time step loop. SIM model execution begins with reading and organizing input data. Water rights are sorted into priority order based on priority numbers and/or other user-defined options. The simulation steps through time. Naturalized flows for primary control points and net evaporation rates for reservoirs are read at the beginning of the time step loop. Flows are distributed from primary control points to all other sites based on watershed parameters. Within each sequential month, water accounting computations are performed as each set of water use requirements (water right) is considered in priority order.

Water allocation and management are modeled by accounting procedures within the water rights priority loop. An array is maintained of stream flow available for appropriation at all control points. As each water right is considered in priority order, the following tasks are performed.

1. The diversion, instream flow, or hydropower target is set starting with an annual amount and set of 12 monthly distribution factors provided as input. The target may be further modified as a function of the storage content in any number of specified reservoirs and naturalized, regulated, or unappropriated flow at any control point.
2. The amount of water available to the water right from stream flow is determined based on the available stream flow array considering the control point of the water right and all downstream control points.
3. Water use requirements are met subject to water availability following specified system operating rules. Water accounting computations are performed to determine the diversion, diversion shortage, end-of-month storage, and related quantities. Reservoirs and hydropower plants necessitate an iterative algorithm since evaporation and hydropower releases are a function of both beginning-of-month and end-of-month storage.
4. The available stream flow array is adjusted for that location and all downstream sites to reflect the effects of the water right. Channel loss factors are applied in translating adjustments for stream flow depletions and return flows to flows at downstream sites.

2.5. WRAP-SIM Simulation Results

Simulation results include time series of any of the variables computed in the simulation covering the period-of-analysis. The model-user selects the control points, water rights, and reservoirs for which simulation results are recorded. The 40 variables written to the main output file include but are not limited to:

- naturalized, regulated, and unappropriated flows, stream flow depletions, and return flows for each selected control point
- channel losses and channel loss credits for each selected control point representing the reach below the control point
- storage volume, surface elevation, net evaporation, inflows, releases, diversions, and hydroelectric energy at each selected reservoir
- diversion targets and shortages, return flows, available stream flows, stream flow depletions, and storage for each selected water supply right
- hydropower targets, firm energy produced, secondary energy produced, energy shortages, and storage for each selected hydroelectric power right
- instream flow target and shortage for each selected instream flow right

The WRAP post-simulation program TABLES organizes SIM or SIMD simulation results in various user-specified formats including time series of selected variables, water budgets, statistical summaries, and various types of frequency relationships and reliability indices. Tables may be created in a format for incorporation in reports or in formats convenient for export to Microsoft Excel or HEC-DSSVue [20].

2.6. Daily WRAP Modeling System

WRAP and the Texas WAM system employ a monthly step time. However, recent versions of WRAP also include capabilities for daily simulations [18]. The daily WRAP is applicable to all aspects of river/reservoir system management. However, its development has been motivated largely by the need for expanded capabilities for modeling environmental flow requirements and issues [22]. The daily WRAP modeling system is being applied with the Texas WAM system to supplement rather than replace the monthly models. Alternative strategies can be adopted in applying monthly and daily simulations in combination. Both monthly and daily simulations may be applied with the results of each used to support particular decision processes. Alternatively, daily environmental instream flow targets computed in a daily simulation can also be aggregated to monthly target series for input to a monthly simulation, using WRAP features that conveniently facilitate this modeling strategy.

The WRAP daily modeling system includes the following additional features used only with a daily computational time step:

- disaggregation of monthly naturalized flows to daily
- flow routing and forecasting
- disaggregation of diversion, hydropower, and instream flow targets
- simulation of pulse flow environmental flow requirements
- simulation of reservoir flood control operations
- additional frequency analysis capabilities

Fundamental differences between daily and monthly simulations are as follows.

1. Flow rates that vary continuously over time in the real world are modeled as volumes occurring during discrete time intervals. The smaller computational time interval improves the accuracy of the simulation and is particularly relevant in simulating environmental flow requirements and flood control operations. Effects of a smaller computational time step are diminished by reservoir storage and tend to be least pronounced for water supply rights with large reservoir storage capacities.
2. In a monthly model, effects of reservoir releases and water management/use actions on stream flows at downstream locations are assumed to propagate through the river system within the same month, precluding flow forecasting and routing computations. However, flow forecasting and routing are important with a daily time step.
3. Reservoir flood control operations can be incorporated in a daily model.
4. Environmental high pulse flows can be incorporated in a daily model.

Daily naturalized flows input to SIMD may be used directly in the simulation, or alternatively SIMD can disaggregate monthly naturalized flows based on patterns defined by inputted daily flows while maintaining the original monthly volumes. WRAP provides flexible options for making the best use of limited available sequences of daily flows, allowing flows at selected locations to serve as pattern hydrographs for disaggregating flows at multiple other sites. Daily naturalized flows have been developed by adjusting gaged flows to remove the effects of regulation and water use. In some cases for relatively unregulated tributary streams, observed flows have been adopted for use as pattern hydrographs without adjustments. Daily flow patterns can also be established with watershed precipitation-runoff models.

A lag and attenuation routing method was developed specifically for the WRAP daily simulation model SIMD. Water supply diversions, return flows, reservoir releases and storage refilling result in changes in stream flows at downstream locations. Routing in SIMD refers to the downstream propagation of changes resulting from an upstream change to stream flow. Two parameters called the lag factor and attenuation factor are determined with a calibration procedure for each specified routing reach based on known daily flow sequences at the upstream and downstream ends. The lag and attenuation parameters are input to SIMD for use in propagating flow changes downstream. A reverse routing algorithm is also applied to replicate the effects of routing in the procedure for forecasting flow availability.

Flow forecasting makes the internal daily computations in SIMD much more complicated than a monthly simulation. River flows diverted or stored by a particular water user today may diminish flows available to other downstream water users in future days. Likewise, flow travel times for reservoir releases or diversion return flows to reach downstream locations may be several days. Thus, water supply capabilities are affected by earlier upstream activities. Likewise, flood control reservoir operations are based on making no releases that contribute to flows exceeding maximum non-damaging flow limits at downstream gages that may be located several days of flow travel time below the dam. The flow forecasting algorithm in SIMD is based on a forecast simulation used solely to derive future flow availability information. For each day of the simulation, the final simulation for that day is preceded by a simulation covering a future forecast period that generates stream flow availability information for that current day.

SIMD simulates flood control operations of any number of reservoirs based on allowable flows at any number of downstream control points. Flood control pools are emptied as quickly as feasible subject to making no releases that contribute to flows at downstream control points exceeding allowable flow limits. Storage volume versus maximum outflow relationships at each individual reservoir are also specified.

2.7. Simulation Modes

In the conventional long-term simulation mode, a specified water management/use scenario is combined with naturalized flows and net reservoir evaporation rates covering the entire hydrologic period-of-analysis in a single simulation. The example discussed later in this chapter has a 1940-2012 hydrologic period-of-analysis. In the conditional reliability modeling mode, discussed later, the period-of-analysis hydrology is divided into many sequences, and the simulation is automatically repeated with each hydrologic sequence starting with the same specified initial storage condition. In the example, the1940-2012 period-of-analysis is divided into 73 annual hydrologic simulation sequences.

The simulation model also has options that involve repetitions of the complete long-term hydrologic simulation period. With the yield-reliability option, one or more selected diversion rights start with a specified target that is iteratively incremented until the firm yield is reached. A dual simulation option is useful in modeling multiple rights with different priorities associated with the same reservoir system. Another option sets reservoir storage contents at the beginning of a second simulation equal to the storage at the end of an initial simulation.

3. Texas Water Availability Modeling System

Texas is a large state located in the south-central United States that is representative of both the drier western and wetter eastern regions of the country from the perspectives of climate and water management. Texas has an area of 685,000 km^2 (264,000 square miles). Climate, geography, and water management vary dramatically across the state from the arid western desert to humid eastern forests, from sparsely populated rural regions in the western and eastern extremes of the state to the metropolitan areas of Dallas and Fort Worth, Austin, San Antonio, and Houston shown in Figure 1. Mean annual precipitation varies from 20 cm (8.0 inches) at El Paso on the Rio Grande to 142 cm (56 inches) in the lower Sabine River Basin. Population and economic growth combined with depleting groundwater reserves have resulted in increasing demands on surface water resources throughout the state. Environmental flow requirements and ecosystem preservation are major concerns as well as meeting municipal, industrial, and agricultural water needs. The population of the state increased from 5.82 million people in 1930 to 25.4 million in 2010 and is projected to increase to 46.3 million by 2060 [23].

Figure 1. Map of Texas with major rivers, largest cities, and neighboring states.

The Texas Commission on Environmental Quality (TCEQ) as lead agency, Texas Water Development Board (TWDB), Texas Parks and Wildlife Department (TPWD), and their contractors, consisting of two universities and ten consulting engineering firms, implemented the WAM System pursuant to comprehensive water management legislation enacted by the Texas Legislature in 1997 [24]. The modeling system is routinely applied by applicants, or their consultants, in preparation of water right permit applications and by TCEQ staff in evaluating the applications. The TWDB is the lead agency for regional and statewide planning studies that represent another major application of the WAM System. River authorities and other water management entities also apply the WAM System in endeavors not directly mandated by the TCEQ water rights permitting or TWDB planning programs. The WAM System supports a broad range of water management activities and helps to integrate those activities.

Table 2. River Basin Models in the Texas WAM System

Major River Basin or Coastal Basin	Area In Texas (km^2)	Area Outside Texas (km^2)	Number of			Reservoir Storage Capacity (10^6 m^3)
			Control Points	Water Rights	Reservoirs	
Brazos River	115,000	6,660	3,830	1,634	670	5,758
Canadian River	32,900	90,700	85	56	47	1,192
Colorado River	108,000	5,100	2,395	1,922	511	5,878
Cypress Bayou	7,280	259	189	164	91	1,078
Guadalupe-San Antonio	26,500	-0-	1,349	860	237	997
Lavaca River	5,980	-0-	185	71	22	290
Neches River	25,900	-0-	318	333	176	4,818
Nueces River	43,900	-0-	542	373	121	1,284
Red River	63,400	61,000	447	489	245	4,965
Sabine River	19,200	6,040	376	310	207	7,873
San Jacinto River	14,500	-0-	411	148	114	787
Sulphur River	9,220	492	83	85	53	930
Trinity River	46,500	-0-	1,334	1,169	703	9,254
Rio Grande	125,000	347,000	957	2,584	113	16,149
Coastal Basins						
Lavaca-Guadalupe	2,590	-0-	68	10	-0-	-0-
Neches-Trinity	1,990	-0-	245	138	31	40
Lower Nueces-Rio Gr.	17,000	-0-	119	70	42	90
Upper Nueces-Rio Gr.	10,000	-0-	81	34	22	50
San Antonio-Nueces	6,860	-0-	53	12	9	2
Trinity-San Antonio	648	-0-	94	24	13	6
Colorado-Lavaca	2,440	-0-	111	27	8	67

WRAP is generalized for application to river/reservoir systems located anywhere in the world, with model-users developing input datasets for the particular river basins of concern. In Texas, TCEQ WAM System WRAP input datasets are altered as appropriate to reflect proposed water management plans of interest, which may involve changes in water use or reservoir/river system operating practices, construction of new facilities, or other water management strategies. The 21 input datasets covering 15 major river basins and eight coastal basins are listed in Table 2. The datasets are available at the previously cited TCEQ website. The water rights in the datasets are updated as the TCEQ approves applications for new permits or revisions to existing permits. The datasets also continue to be refined with addition of expanded WRAP capabilities. Incorporation of expanded environmental flow standards is currently a major focus.

The datasets model the approximately 3,435 reservoirs for which a water right permit has been issued. Over 90 percent of the total capacity of the 3,435 reservoirs is contained in the 210 reservoirs that have conservation capacities exceeding 6.17 million m^3 (5,000 acre-feet). In developing the WAM datasets, elevation-storage-area tables for most of the major reservoirs having conservation storage capacities of at least 6.17 million m^3 (5,000 acre-feet) were assembled for both permitted and estimated year 2000 conditions of sedimentation. Generalized storage-area relationships were adopted in each river basin for the numerous smaller reservoirs.

Several of the Texas river systems are shared with neighboring states. The Rio Grande is shared with Mexico. For the interstate and international river basins, hydrology and water management in neighboring states and Mexico are considered to the extent necessary to assess water availability in Texas. The models reflect two international treaties and five interstate compacts as well as two different Texas water rights systems administered by the TCEQ.

The WRAP input datasets in the WAM system include alternative scenarios reflecting combinations of premises regarding water use, return flows, and reservoir sedimentation. The following current use and authorized use scenarios described below are routinely adopted for water right permit applications, planning studies, and other applications. The TWDB has also developed alternative future use scenarios for different points in time over a 50 year future planning horizon.

The authorized use scenario WRAP input datasets model the over 6,000 water right permits as written. Water supply diversion targets are based on the premise that all permit holders use the full amounts to which they are legally entitled. Reservoir storage capacities are based on data from the permits which typically does not include adjustments for reservoir sedimentation. Return flows for permitted diversions are included only if return flows are required by the permits. Temporary short-term permits are not included in the datasets.

The current use scenario WRAP input datasets incorporate adjustments to reflect actual current conditions of water resources development, allocation, and use. Water users often use less than their permits allow. The current use scenario incorporates the maximum actual annual diversion observed in recent years for each individual permit. Best estimates of return flows are incorporated in the model. Reservoir Storage capacities are adjusted to reflect estimated year 2000 or later conditions of sedimentation. Temporary short-term permits are included.

Major reaches of rivers, particularly in dry west Texas and populous urban areas elsewhere in the state, are over-appropriated. Reliabilities for supplying permitted water use are unacceptably low, and the TCEQ will not issue permits for additional use. Marketing or transferring of existing water rights among users is encouraged. For other river reaches, water is available for further appropriation. The TCEQ continues to issue or modify many water right permits each year. Additional new environmental flow standards appropriating additional quantities of stream flow are currently being incorporated in the WAM system.

The WAM System is routinely applied by water management agencies and their consultants in planning studies and preparation of permit applications. TCEQ staff applies the modeling system in evaluating the permit applications. The TWDB, local regional planning groups, and consulting engineering firms, apply the modeling system in regional planning studies which are formally updated in five year cycles. The regional plans are incorporated into a statewide plan. Water right permit applications must be consistent with relevant regional and statewide plans. Agencies and engineering firms use the modeling system in various other types of studies as well.

4. Example River/Reservoir System Simulation

The WRAP Fundamentals Manual [15] is organized around an example which is adopted here to illustrate the concepts presented in this chapter. Examples in the Reference, Salinity, and Daily Manuals [13, 17, 18] build upon and extend the example in the Fundamentals Manual. The SIM, SIMD, SALT, and TABLES input files for all of the examples in the manuals are available at the WRAP website along with the manuals and executable programs. The basic simulation input dataset for the example adopted here is described in detail in the Fundamentals Manual [15].

The example presented in the Fundamentals Manual was adapted from the TCEQ WAM system dataset for the Brazos River Basin (Table 1) which has 670 reservoirs, 77 primary control points, and over 3,000 secondary control points [22, 24]. Over a thousand water districts, cities, companies, and individuals hold permits to use the waters of the Brazos River and its tributaries. The example is reduced to a system of six reservoirs and eleven control points shown in Figure 2 and Table 3 along with 30 water rights representing a hypothetical present condition scenario that includes two environmental instream flow rights, hydroelectric energy generation at Whitney Reservoir, and 26 diversion rights at the 11 control points for municipal, industrial, and agricultural water supply. The 11 control points listed in Table 3 include the sites of six dams and five stream gauging stations and also represent diversion sites in the model.

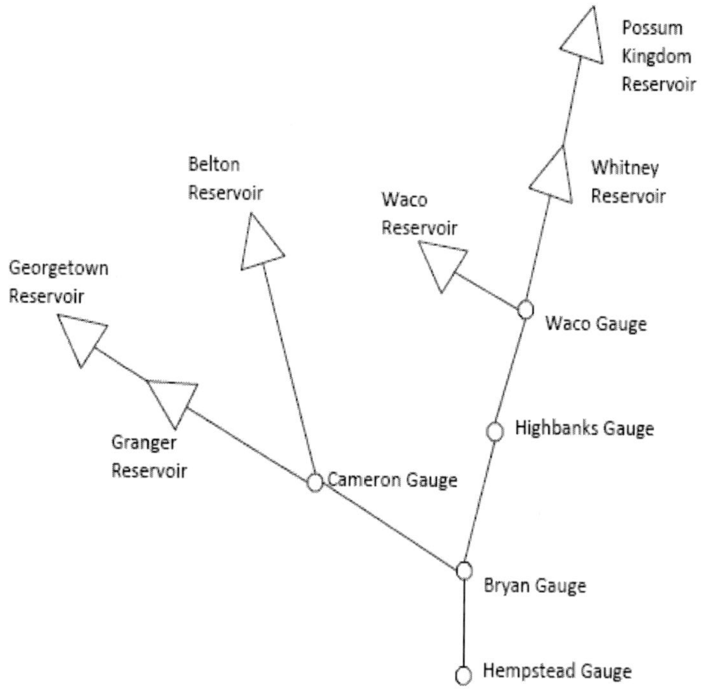

Figure 2. System Schematic for the Example.

Either metric or English units can be used in WRAP. However, applications in Texas and the examples in the manuals use English units. English units are adopted for the examples of this chapter to facilitate comparison with the examples in the manuals. An acre-foot is a volume equivalent to covering an area of one acre to a depth of one foot and equals 1,234 cubic meters.

Possum Kingdom Reservoir is owned and operated by the Brazos River Authority (BRA). The other five federal multiple-purpose reservoirs are owned and operated for flood control by the U.S. Army Corps of Engineers (USACE). The BRA has contracted for the conservation storage capacity of the five federal reservoirs and operates the six-reservoir system for water supply. The USACE/BRA system contains three other reservoirs that are not included in the example.

Table 3. Example River/Reservoir/Use System

Control Point	River	Watershed Drainage Area (sq. mile)	Mean Naturalized Flow (ac-ft/yr)	Conservation Storage Capacity (acre-feet)	Annual Diversion Demands (ac-ft/yr)
Possum Kingdom Dam	Brazos River	14,100	63,290	570,240	254,800
Whitney Dam	Brazos River	17,700	112,200	627,100	18,000
Waco Dam	Bosque River	1,660	31,280	192,100	80,800
Belton Dam	Leon River	3,580	37,670	457,100	180,260
Georgetown Dam	San Gabriel	404	4,950	37,100	25,610
Granger Dam	San Gabriel	737	16,250	65,500	42,000
Cameron Gauge	Little River	7,100	112,600	-0-	209,600
Waco Gauge	Brazos River	20,100	156,900	-0-	32,300
Highbanks Gauge	Brazos River	20,900	187,000	-0-	44,800
Bryan Gauge	Brazos River	30,000	329,200	-0-	98,900
Hempstead Gauge	Brazos River	34,400	440,900	-0-	1,119,700

square mile = 2.59 square kilometers; acre-foot = 1,234 cubic meters.

The reservoir storage capacities and surface elevation versus storage volume and surface area tables are from the Brazos WAM. Conservation storage capacities totaling 1,949,140 acre-feet in the six large reservoirs included in the example account for 42 percent of the total conservation capacity of the 670 reservoirs in the Brazos River Basin WAM dataset.

Water supply diversion targets at the five downstream control points (gauging stations) are supplied by unregulated flows entering the river system below the dams supplemented as necessary by releases from the dams. Multiple-reservoir release decisions in the model are based on balancing storage depletions in the reservoirs expressed as a percentage of conservation storage capacity.

Hypothetical water supply, instream flow, and hydropower demands were formulated for the Fundamentals Manual example to simplify the SIM water rights input dataset. The totals of annual diversion targets at each control point are tabulated as the last column of Table 3. The distribution of annual diversion targets over the 12 months of the year vary for the different municipal, industrial, and agricultural water rights. Most of the conservation storage capacity of Whitney Reservoir is allocated to hydroelectric power plant operations with an annual target of 36,000 kilowatt-hours/year that varies seasonally over the 12 months of the year.

The hydrologic period-of-analysis is 1940-2012. The hydrology input data for the simulation model SIM contains the 1940-2012 monthly naturalized flows at the six dams and five downstream gauges and 1940-2012 monthly net water surface evaporation less precipitation rates for each of the six reservoirs from the actual Brazos River Basin WAM dataset.

The monthly naturalized flows at the Hempstead gauge control point plotted in Figure 3 illustrate the great variability of flows of the Brazos River and throughout Texas. Annual totals of the naturalized, regulated, and unappropriated flows are compared in Figure 4. Regulated and unappropriated flows are significantly less than naturalized flows. Unappropriated flows are a little smaller than regulated flows due to environmental instream flow requirements at the Hempstead gauge control point. End-of-month storage contents of the six reservoirs are plotted in Figures 5 and 6. The most severe drought begins during 1950 and ends in April 1957 with the most severe flood occurring at the site during the 1940-2012 hydrologic period-of-analysis. The 1950-1957 drought is the most severe on record statewide. For over half of Texas, including most of the Brazos River Basin, in terms of annual precipitation, 2011 was the driest single year since precipitation records began in 1895. Many other major droughts are evident in the plots.

Setting the initial (beginning of January 1940 in this example) reservoir storage contents in the simulation model is an issue. WRAP provides flexible options for setting the beginning-of-simulation storage contents, including a cycling option for setting the beginning-of-simulation storage contents of each reservoir equal to its end-of-simulation storage contents. All of the Texas WAM system datasets start the simulation with the reservoirs full to capacity and have a hydrologic simulation beginning during a wet period, typically January 1940.

WRAP-SIM simulates

- development, management, allocation, regulation, and use of the water resources of a river basin or multiple-basin region in accordance with water use requirements and institutional water allocation systems utilizing constructed facilities including reservoir projects, conveyance systems, and hydroelectric power plants
- during each sequential month of a hydrologic period-of-analysis with river basin hydrology represented by sequences of monthly

naturalized stream flows and reservoir net evaporation-precipitation depths.

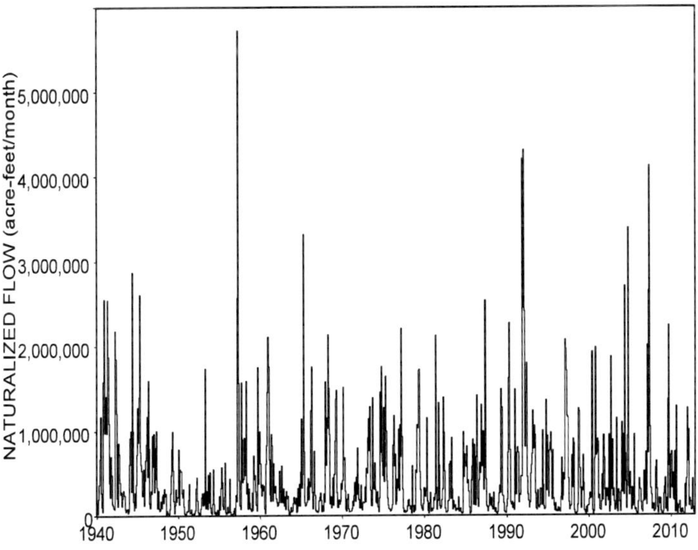

Figure 3. Monthly Naturalized Flow at Hempstead Gauge.

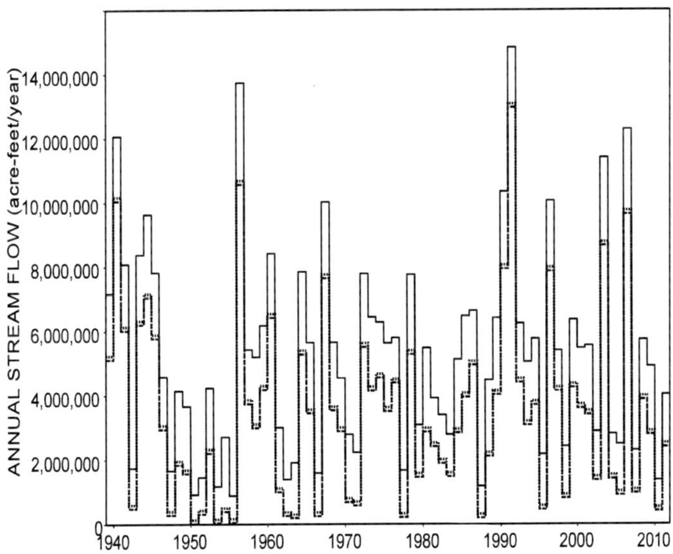

Figure 4. Annual Naturalized (solid line) , Regulated (dotted line), and Unappropriated (dashed line) Flows at Hempstead Gauge.

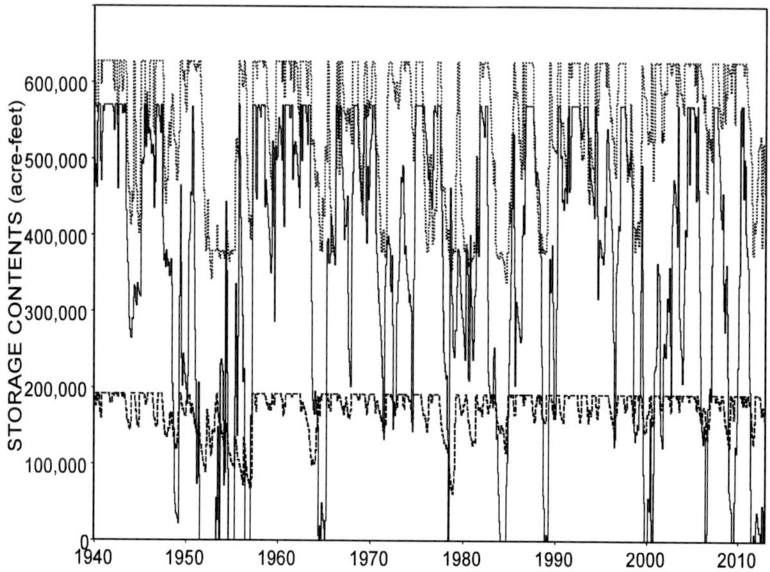

Figure 5. Storage Contents of Lakes Whitney (capacity 627,100 acre-feet), Possum Kingdom (570,240 ac-ft), and Waco (192,100 ac-ft).

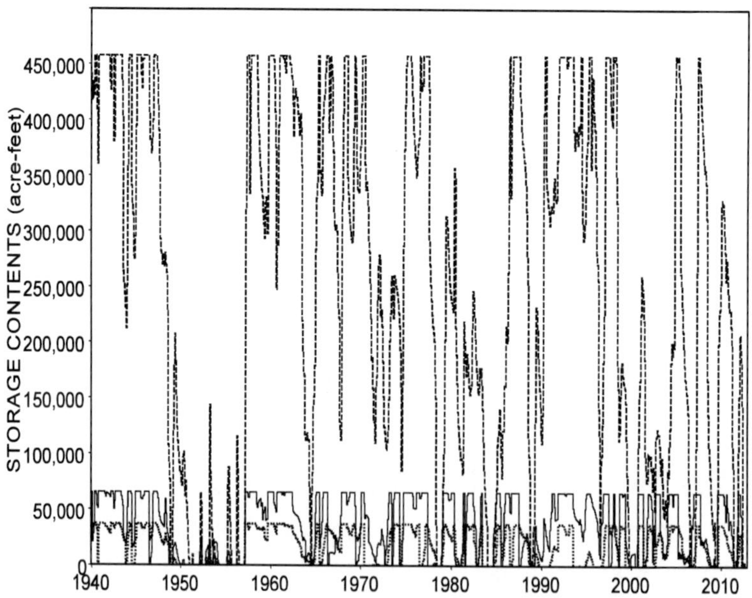

Figure 6. Storage Contents of Lakes Belton (capacity 457,600 acre-feet), Granger (65,500 ac-ft), and Georgetown (37,100 ac-ft).

Capabilities are assessed for meeting water use requirements under a defined scenario of water resources development and management during given sequences of naturalized flows and net reservoir evaporation-precipitation rates. The future is of concern, not the past. However, since future hydrology is unknown, historical past stream flows and reservoir evaporation-precipitation rates are adopted as being statistically representative of the hydrologic characteristics of a river basin that can be expected to continue into the future.

The results of a *SIM* simulation are viewed from the perspectives of likelihood, frequency, probability, percent-of-time, or reliability of meeting water supply, instream flow, hydropower, and reservoir storage requirements. Options incorporated in the WRAP program TABLES include the following metrics for summarizing SIM, SIMD, and SALT simulation results.

- frequency tables for naturalized, regulated, or unappropriated flows, reservoir storage volumes or surface elevations, or any of the other time series variables in the SIM or SIMD simulation results or computed within TABLES from the simulation results
- reliability tables for water supply diversion or hydroelectric energy targets
- diversion shortage summary metrics
- flood frequency analyses of peak annual flows or storage from a SIMD simulation
- frequency tables for salinity loads or concentrations from SALT simulation results

5. Frequency Analysis Metrics

Exceedance frequency metrics express the percentage-of-time or equivalently the likelihood or probability that a variable of interest will equal or exceed a certain amount. Exceedance frequency tables may be created with TABLES for essentially any of the time series variables included in the simulation results and other variables computed within TABLES from the simulation results. Key variables for which frequency analyses are performed include: naturalized flow, regulated flow, unappropriated flow, or reservoir storage volume for specified control points; reservoir storage or drawdown volume or water surface elevation for specified reservoirs; and instream flow targets or shortages for specified water rights or control points.

The monthly simulation model SIM generates time series with quantities for each month of the hydrologic period-of-analysis. The daily simulation model SIMD produces quantities for each day of the hydrologic period-of-analysis. The post-simulation program TABLES also produces annual series from monthly or daily series that may contain the total, mean, minimum, or maximum value for each year. Optionally, frequency analyses may be applied in TABLES to moving averages or moving accumulative totals of the data computed for a user-specified number of months or days. For example, an annual series of the minimum 7-day flow volume in each year can be produced from a daily series of naturalized, regulated, or unappropriated flows. Frequency metrics may reflect all months or alternatively be developed for a specified month of the year such as May or August. TABLES include flexible options for creating frequency tables in various alternative formats and applying alternative computation methods.

5.1. Relative Frequency

Exceedance frequencies may be determined from the simulation results based on counting the relative frequency in which various quantities are equaled or exceeded. Relative frequency is expressed by Equation (1) where n is the number of months or days during the simulation that a particular flow or storage amount is equaled or exceeded, and N is the total number of months or days considered.

$$\text{Exceedance Frequency} = \frac{n}{N} \, (100\%) \tag{1}$$

The example is based on a 73-year 1940–2012 hydrologic period-of-analysis and monthly computational time step. N is 876 months with all months considered in the frequency analysis. N is 73 months if only a selected month, such as July, is considered.

Stream flow volumes for each of the 876 months of the simulation are plotted in Figure 3. Simulated end-of-month storage volumes for each of the 876 months are plotted in Figures 5 and 6. Tables 4 and 5 were developed by applying Eq. (1) to the 876 months of the simulation. The simulation could be repeated with a daily computational time step resulting in an N in Eq. (1) of

26,665 days and the variables being daily flow volumes or end-of-day storage volumes.

The mean, standard deviation, and quantities for specified exceedance frequencies are tabulated for each of the six reservoirs in Table 4. The last column is derived by applying Eq. (1) to the total storage in the six reservoirs at the end of each of the 876 months of the simulation. The storage volume totals associated with exceedance frequencies of 75%, 50% (median), and 25% are 1,036,820 acre-feet, 1,445,480 acre-feet, and 1,756,070 acre-feet, respectively.

Frequency metrics for naturalized, regulated, and unappropriated flows at the Cameron gauge and Richmond gauge control points are compared in Table 5. The mean and median (50% frequency) naturalized flow at the Cameron control point are 112,620 and 43,533 acre-feet/month, respectively, for the 876 months of the 1940-2012 period-of-analysis.

The 1940-2012 historical hydrology is past and will not be repeated in the future. However, the modeling strategy is based on the premise that the frequency metrics will be reasonably representative of the future as well as the past. Interpretation of the frequency metrics is illustrated as follows.

Table 4. Frequency Metrics For Reservoir Storage Contents (acre-feet)

Reservoir	Possum K	Whitney	Waco	Belton	George	Granger	Total
Mean	353,929	540,659	172,056	249,828	18,376	36,108	1,370,955
Stand Dev	187,894	90,379	26,927	164,738	15,398	26,734	438,913
End-of-Month Storage Contents (acre-feet) for Specified Exceedance Frequencies							
100%	0	337,714	61,608	0	0	0	452,414
99%	0	360,674	77,872	0	0	0	478,036
98%	0	369,065	91,339	0	0	0	495,149
95%	0	379,000	113,216	0	0	0	572,884
90%	27,968	379,000	133,888	0	0	0	707,415
85%	128,767	411,366	146,425	35,181	0	0	839,977
80%	168,519	443,409	154,896	76,370	0	3,322	928,523
75%	214,651	472,728	161,005	101,181	0	8,308	1,036,825
70%	253,992	498,722	166,553	121,265	2,184	11,707	1,110,579
60%	330,390	530,514	175,508	186,579	8,980	22,888	1,296,353
50%	387,864	568,240	183,130	256,423	19,773	40,212	1,445,477
40%	446,955	602,333	190,927	329,159	27,733	54,394	1,551,814
30%	501,159	627,100	192,100	389,302	33,044	65,344	1,688,014
25%	532,693	627,100	192,100	420,134	35,067	65,500	1,756,066
20%	554,966	627,100	192,100	445,734	36,845	65,500	1,826,012
15%	570,240	627,100	192,100	457,600	37,100	65,500	1,894,677
10%	570,240	627,100	192,100	457,600	37,100	65,500	1,936,533
Maximum	570,240	627,100	192,100	457,600	37,100	65,500	1,949,640

Table 5. Frequency Metrics for Stream Flow (acre-feet/month)

	Flow at Cameron Gauge			Flow at Hempstead Gauge		
	Naturalized	Regulated	Unappropr	Naturalized	Regulated	Unappropr
Mean	112,620	85,727	74,826	440,872	294,084	284,084
Stand Dev	180,174	156,303	156,508	591,954	536,032	536,043
Monthly Flow Volumes (acre-feet/month) for Specified Exceedance Frequencies)						
100%	0	0	0	1,634	9,205	0
99%	277	296	0	12,661	9,205	0
98%	1,026	296	0	16,471	9,863	0
95%	2,157	296	0	26,175	9,863	0
90%	4,779	306	0	39,938	9,863	0
85%	8,348	306	0	55,072	9,863	0
80%	11,394	2,392	0	69,580	10,192	0
75%	15,023	5,226	0	84,472	10,192	0
70%	18,475	8,141	0	109,106	10,192	0
60%	28,550	14,818	0	156,315	14,128	4,411
50%	43,533	26,844	9,382	220,827	56,913	47,050
40%	61,439	45,489	22,304	302,189	122,218	112,224
30%	100,182	74,215	48,128	448,663	254,713	244,521
25%	129,579	93,225	73,064	576,601	381,521	371,658
20%	171,099	118,027	106,946	721,320	499,544	489,549
15%	233,242	160,731	159,086	895,000	658,730	648,735
10%	303,417	226,039	222,583	1,141,901	861,605	851,611
5%	432,120	386,892	386,594	1,587,840	1,321,882	1,311,690
2%	723,126	611,973	611,672	2,255,012	2,013,376	2,003,184
1%	942,919	761,419	761,113	2,662,897	2,424,054	2,414,112
0.5%	1,212,508	1,100,486	1,017,881	3,843,784	3,500,486	3,490,623
Maximum	1,403,136	1,392,559	1,392,282	5,723,482	5,455,438	5,445,246

A regulated flow of 5,226 acre-feet/month at the Cameron control point is equaled or exceeded during 75 percent of the 876 months of the simulation. Thus, regulated flows equal or exceed an estimated 5,226 acre-feet/month during 75 percent of the time for the water management and use scenario and modeling premises incorporated in the WRAP-SIM simulation model and input dataset. The estimated probability that the flow volume at the Cameron control point will be at least 5,226 acre-feet in any randomly selected past or future month is 0.75 based on the premises reflected in the model.

5.2. Probability Distribution Functions

Relative frequency based on Eq. (1) is routinely applied to perform frequency analyses in applications of the Texas WAM system. Alternatively, TABLES also provides options to apply the normal (Eq. (2)) or log-normal (Eq. (3)) probability distributions to the series of flow, storage, or other quantities generated by SIM, SIMD, or SALT or adjustments thereto.

$$X = \overline{X} + z\,S \tag{2}$$

$$\log X = \overline{\log X} + z\,S_{\log X} \tag{3}$$

$\overline{X}$ and S denote the mean and standard deviation of the selected random variable from the simulation results. The mean and standard deviation of the logarithms of the data are applied in Eq. (3). The frequency factor (z) is derived from a normal probability table. TABLES contain the frequency factors (z) for the normal probability distribution for the exceedance frequencies tabulated in the first column of Tables 4 and 5. The log-normal distribution consists of the normal distribution applied to the logarithms of X, with Eq. (2) expressed as Eq. (3) and z still derived from the normal probability distribution.

The format of the resulting frequency table is the same regardless of whether TABLES uses relative frequency (Eq. (1)) or one of the probability distributions (Eq. (2) or Eq. (3)) to perform the computations. The choice between applying the concept of relative frequency directly versus adopting the normal or log-normal probability distributions depends upon the particular variable and application. When in doubt, the default relative frequency counting option based on Eq. (1) should probably be adopted. The normal or log-normal probability distribution functions offer potential improvements in the accuracy of the frequency estimates. One of the two alternative distributions should be considered for adoption if statistical goodness-of-fit tests demonstrate a reasonably close fit to data indicating that the probability distribution accurately models the physical processes. However, if the probability distribution function is not known to accurately model the relevant phenomenon, Eq. (1) is the optimal choice of computation method.

5.3. Flood Frequency Analysis

The WRAP program TABLES performs flood frequency analyses of annual series derived from daily SIMD simulation results using standard methods described by Wurbs [26] that are based on the log-Pearson type III probability distribution. The log-Pearson type III distribution is applied with Eq. (3) with the frequency factor (z) being a function of the skew coefficient. Frequency factors are incorporated in the TABLES software. The parameters of the probability distribution function computed by TABLES are the mean, standard deviation, and skew coefficient for maximum annual naturalized or regulated flows or reservoir storage volumes.

Alternatively, the series of annual peak daily flow or storage volumes produced by the SIMD simulation may be exported to the HEC-SSP Statistical Software Package [27] available from the Hydrologic Engineering Center (http://www.hec.usace.army.mil/). HEC-SSP provides comprehensive capabilities for performing flood frequency analyses.

6. Reliability Analysis Metrics

The terms "*target*", "*demand*", "*need*", and "*requirement*" are used interchangeable in the following discussion and may refer to either water supply for municipal, industrial, agricultural, or other types of water use or to hydroelectric energy generation. Volume and period reliabilities provide concise metrics for measuring capabilities for meeting water supply diversion and hydroelectric energy generation requirements. Firm yield is the maximum demand that can be supplied with volume and period reliabilities of 100 percent.

6.1. Volume and Period Reliabilities

Volume reliability is the percentage of the total target demand amount that is actually supplied. For water supply diversions, the amounts are volumes. For hydroelectric power, the amounts are kilowatt-hours of energy generated. Volume reliability (R_V) is the ratio of volume of water supplied or the energy produced (v) to the target (V), converted to a percentage.

$$R_V = \frac{v}{V}(100\%) \tag{4}$$

Equivalently, for water supply, R_V is the mean actual diversion rate as a percentage of the mean target diversion rate. For hydropower, R_V is the mean actual rate of energy production as a percentage of the mean target energy production rate.

Period reliability is the percentage of the total number of periods of the simulation during which the specified demand target is either fully supplied or at least a specified percentage of the target is supplied. The various variations of period reliability (R_P) are computed by TABLES from the results of a SIM or SIMD simulation as:

$$R_P = \frac{n}{N} (100\%) \tag{5}$$

where n denotes the number of periods (days, months, years) during the simulation for which the specified percentage of the demand target is met, and N is the total number of periods considered. The standard TABLES reliability table includes metrics with periods defined alternatively in terms of months (or days) and years. N and n can be defined on a monthly basis optionally either considering all months or only months with non-zero demand targets.

Whereas volume reliabilities are based on total diversion volumes or hydroelectric energy amounts, period reliabilities are based on percent-of-time. Period reliability R_P is an expression of the percentage of time that the full demand target or a specified portion of the demand target can be supplied. Equivalently, R_P represents the likelihood or probability of the target being met in any randomly selected month or year. The volume reliability R_V provides a greater focus on the relative magnitude of the shortage amounts.

Water supply diversion reliabilities may be tabulated for individual water rights, the aggregation of all rights associated with individual control points, or user-selected groups of water rights. The WRAP user selects which water rights, control points, or water rights groups to include in a diversion reliability table. The reliabilities in Tables 6, 7, and 8 are for the aggregate of all diversions at each of the eleven control points. Multiple water rights at each of the control points include municipal, industrial, and agricultural water users with different seasonal water use distributions and return flow specifications.

The total aggregated annual diversion requirements at each of the eleven control points are tabulated in the second column of Table 6. The mean total shortages in the third column are the portions of the diversion targets that were not supplied due to insufficient water being available from stream flow and reservoir storage. The period and volume reliabilities in the last two columns of Table 6 are computed with Eqs. (4) and (5). The period reliability is the percentage of the 876 months of the simulation during which water right diversion targets were fully supplied without any shortage. Shortages in supplying the targets vary between months and years of the simulation. The diversion targets at all control points total to 2,106,769 acre-feet/year and the corresponding total shortages average 152,794 acre-feet/year over the 73 year simulation resulting in a volume reliability (Eq. (4)) of 92.75 percent.

Reliability metrics also include tabulations of period reliabilities expressed both as the percentage of months (Table 7) and the percentage of years (Table 8) during the simulation during which diversions or energy produced equaled or exceeded specified magnitudes expressed as a percentage of the target demand. For example, Table 7 shows that at least 94.5 percent of the diversion requirement at the Hempstead gauge control point is supplied during 75 percent of the 876 months of the simulation. Table 8 indicates that at least 93.2 percent of the diversion requirement at the Hempstead gauge control point is supplied during 75 percent of the 73 years of the simulation. Tables 7 and 8 also include columns the percentage of months for which the demand is fully 100% supplied, without any shortage. This 100% column of Table 7 is the same as the period reliabilities in Table 6.

Reliability tables for hydroelectric power have the same format as for water supply diversions. Reliabilities are computed in the example for hydroelectric energy generation at Whitney Reservoir. The monthly hydropower targets sum to a total annual target of 36,000 kilowatt-hours/year. The mean annual shortage is 1,813.5 kilowatt-hours/year, and thus the mean annual energy production from Eq. (3) is 94.96 percent of the target. The energy target is fully met during 92.24 percent (Eq. (4)) of the 876 months of the simulation. Energy production equals or exceeds 75 percent of the target during 93.2 percent of 876 months and also during 93.2 percent of the 73 years of the simulation.

Table 6. Water Supply Diversion Reliabilities

Control Point	Diversion Target (ac-ft/year)	Mean Shortage (ac-ft/year)	Volume Reliability (percent)	Period Reliability (percent)
Possum Kingdom	254,800	11,183	95.61	94.18
Whitney Reservoir	18,000	1,002	94.43	94.63
Waco Reservoir	80,800	0	100.00	100.00
Waco Gauge	32,300	2,657	91.77	95.21
Highbanks Gauge	44,800	5,044	88.74	94.63
Belton Reservoir	180,260	14,367	92.03	90.07
Georgetown Res	25,610	4,675	81.74	77.28
Granger Reservoir	42,000	3,947	90.6	87.21
Cameron Gauge	209,600	37,509	82.1	79.57
Bryan Gauge	98,900	5,793	94.14	88.93
Hempstead Gauge	1,119,699	66,615	94.05	82.53
Total	2,106,769	152,794	92.75	–

Table 7. Monthly Water Supply Diversion Reliabilities

Control Point	Percentage of Months with Diversions Equaling or Exceeding Percentage of Target						
	100%	95%	90%	75%	50%	25%	1%
Possum Kingdom	94.2	94.4	94.5	94.6	95.1	96	98.4
Whitney Reservoir	94.6	94.6	94.6	94.6	94.6	94.7	95.1
Waco Reservoir	100.0	100.0	100.0	100.0	100.0	100.0	100.0
Waco Gauge	95.2	95.4	95.5	95.9	96.5	96.6	97.0
Highbanks Gauge	94.6	94.6	94.6	95.1	95.8	96.3	96.8
Belton Reservoir	90.1	90.3	90.3	90.8	91.2	92.9	95.5
Georgetown Res	77.3	77.5	78	79.3	81.5	83.9	93.2
Granger Reservoir	87.2	87.4	87.9	88.7	91.0	92.7	96.2
Cameron Gauge	79.6	80.5	80.9	83.7	89.4	95.8	99.3
Bryan Gauge	88.9	89	89.4	93.8	98.7	99.3	99.7
Hempstead Gauge	82.5	85.6	87.7	94.5	96.3	97.7	99.8

Table 8. Annual Water Supply Diversion Reliabilities

Control Point	Percentage of Years with Diversions Equaling or Exceeding Percentage of Target						
	100%	98%	95%	90%	75%	50%	1%
Possum Kingdom	78.1	79.5	79.5	84.9	95.9	97.3	100.0
Whitney Reservoir	83.6	83.6	83.6	84.9	90.4	95.9	100.0
Waco Reservoir	100.0	100.0	100.0	100.0	100.0	100.0	100.0
Waco Gauge	60.3	64.4	67.1	74.0	79.5	97.3	100.0
Highbanks Gauge	54.8	56.2	57.5	64.4	75.3	94.5	100.0
Belton Reservoir	74.0	75.3	76.7	82.2	87.7	91.8	100.0
Georgetown Res	49.3	52.1	56.2	60.3	69.9	82.2	100.0
Granger Reservoir	61.6	64.4	67.1	76.7	86.3	93.2	100.0
Cameron Gauge	12.3	13.7	21.9	37.0	72.6	93.2	100.0
Bryan Gauge	39.7	43.8	60.3	74.0	95.9	100.0	100.0
Hempstead Gauge	26.0	45.2	71.2	83.6	93.2	98.6	100.0

Reliability estimates reflect all of the premises and approximations incorporated in the modeling system and thus are necessarily approximate. Hydrologic variables such as river flows are highly variable, stochastic, and uncertain. The model and resulting reliabilities are based on the premise that historical hydrology is statistically representative of the climatic and hydrologic characteristics of the river basin to be expected in the future. Data describing the physical and institutional features of water management systems are also necessarily approximate. Uncertainties are inherent in modeling water management and use practices.

The reliability statistics may be interpreted as follows. The water supply diversions at the Hempstead gauge control point equal or exceed 90 percent of the monthly target an estimated 87.7 percent of the time as shown in Table 7. At least 90 percent of the annual demand is supplied an estimated 83.6 percent of the time as shown in Table 8. For the water resources development, allocation, and use conditions reflected in the model, the estimated probability of supplying at least 90 percent of the monthly target at the Hempstead gauge control point in any randomly selected future month is 0.877. The estimated probability of supplying at least 90 percent of the annual target in any randomly selected future year is 0.836.

In evaluating water right permit applications for agricultural irrigation, the Texas Commission on Environmental Quality (TCEQ) approval requires that at least 75 percent of the demand is supplied at least 75 percent of the time, with the period reliabilities computed based both on months and years. TCEQ approval of a municipal water right permit application is based on the criterion of 100.0 percent volume and period reliability. The prior appropriation water rights permit system is based on the concept that new water rights permit applications are not approved if the reliabilities of existing water rights are diminished. The WAM system is based on curtailing junior rights as necessary to protect reliabilities of senior water rights.

Water supply capabilities may depend on water quality as well as quantity. The WRAP programs SALT and TABLES include features for computing reliabilities constrained by specified maximum allowable salinity concentrations [17, 28].

6.2. Firm Yield

Firm yield is a metric that is routinely adopted in planning studies in Texas and elsewhere. The firm yield associated with a water supply diversion or hydroelectric power production target is defined as the maximum annual demand target that can be met with a volume reliability of 100.0 percent. Firm yields may be determined by iteratively adjusting an annual target amount, called the annual yield, and rerunning the simulation model until a yield with zero shortage is found. The WRAP simulation model SIM or SIMD has an option that automates this procedure. The automated yield-reliability routine in SIM and SIMD computes reliabilities with Eqs. (4) and (5) with the target incremented multiple times in iterative repetitions of the simulation.

The iterative SIM routine develops a yield versus reliability table. The firm yield is the last entry in the yield-reliability table, assuming a non-zero firm yield is possible. In some situations, any non-zero target results in a reliability of less than 100 percent. The yield-reliability analysis computations are based on an automated repetition of the entire simulation. One or more selected water rights start with a user-specified annual diversion or hydropower target greater than the firm yield that is iteratively reduced until either the firm yield or a zero target is reached.

An annual water supply diversion target or hydroelectric energy generation target is entered as SIM input. The reliabilities associated with various yields, represented by the annual target amount, may be developed simply by running SIM and TABLES multiple times, manually changing the target entry in the input file for each run. With the automated procedure activated, in a single SIM execution, multiple simulations are repeated with the annual target amount being systematically changed. The resulting yield-reliability table is written as a SIM output file.

The water supply diversion or hydropower yield may be associated with a single water right or a set of any number of water rights. If associated with multiple water rights, two options are available for allocating the annual yield amount between rights. With the first option, the yield is allocated between rights in proportion to target amounts entered in the input. The second option is based on the water right priorities. The yield is assigned to the most senior priority right up to its target amount. Any yield remaining is assigned to the next most senior priority right up to its target amount, and so forth.

A yield-reliability table generated with SIM with the example dataset is presented as Table 9. A diversion right at the Hempstead gauge control point is replaced with a diversion for which the yield-reliability table is developed. The diversion is supplied by unregulated flows entering the river system below the dams supplemented as necessary with releases from the six reservoirs. The diversion has a priority that is junior to all other diversion and instream flow requirements and senior to refilling storage in the six reservoirs.

As indicated by Table 9, the diversion target is set at 1,200,000 acre-feet/year in the initial simulation. The mean shortage of 97,852 acre-feet/year results in a volume reliability of 91.85 percent. The diversion target is reduced by 100,000 acre-feet/year in each subsequent simulation until the mean shortage drops below the set criterion of 0.05 acre-feet/year, which produces a volume reliability of 100.00. This is achieved with an annual diversion target of 300,000 acre-feet/year. The process is automatically repeated beginning with an initial target of 390,000 acre-feet/year and incremental decrease of

10,000 acre-feet/year. The firm yield of 324,800 acre-feet/year is found in the 26th simulation shown in the last line of Table 9.

Each of the 26 iterations reflected in the yield-reliability table is a complete 1940-2012 hydrologic period-of-analysis SIM simulation with all input the same except for the one selected diversion target. Iterations for each of the four levels of target increments stop when the mean annual shortage dropped below 0.05 acre-feet/year. The increment for the 4th level of target

Table 9. Yield-Reliability Table for Selected Diversion Ending with Firm Yield

Iteration	Level	Diversion Target (ac-ft/year)	Mean Shortage (ac-ft/year)	Mean Diversion (ac-ft/year)	Volume Reliability (%)	Months Without Shortage	Period Reliability (%)
1	0	1,200,000	97,852	1,102,148	91.85	783	89.38
2	1	1,100,000	73,143	1,026,857	93.35	796	90.87
3	1	1,000,000	52,564	947,436	94.74	815	93.04
4	1	900,000	35,298	864,702	96.08	829	94.63
5	1	800,000	24,168	775,832	96.98	841	96.00
6	1	700,000	16,219	683,781	97.68	849	96.92
7	1	600,000	9,088	590,913	98.49	860	98.17
8	1	500,000	4,154	495,846	99.17	866	98.86
9	1	400,000	990	399,010	99.75	871	99.43
10	1	300,000	0.00	300,000	100.00	876	100.00
11	2	390,000	1,215.10	388,785	99.69	873	99.66
12	2	380,000	954.50	379,046	99.75	873	99.66
13	2	370,000	716.80	369,283	99.81	874	99.77
14	2	360,000	486.90	359,513	99.86	874	99.77
15	2	350,000	244.90	349,755	99.93	875	99.89
16	2	340,000	137.40	339,863	99.96	875	99.89
17	2	330,000	41.40	329,959	99.99	875	99.89
18	2	320,000	0.00	320,000	100.00	876	100.00
19	3	329,000	33.50	328,967	99.99	875	99.89
20	3	328,000	25.40	327,975	99.99	875	99.89
21	3	327,000	17.40	326,983	99.99	875	99.89
22	3	326,000	9.41	325,991	100.00	875	99.89
23	3	325,000	1.25	324,999	100.00	875	99.89
24	3	324,000	0.00	324,000	100.00	876	100.00
25	4	324,900	0.32	324,900	100.00	875	99.89
26	4	324,800	0.00	324,800	100.00	876	100.00

reductions is 100 acre-feet/year. The simulation results shown in Table 9 indicate that a diversion target of 324,900 acre-feet/year results in a mean shortage of 0.41 ac-ft/yr which exceeds the criterion limit of 0.05 ac-ft/yr. However, the mean shortage for a target of 324,800 ac-ft/yr is below this limit. The firm yield is concluded to be 324,800 acre-feet/year. The last two columns of the table provide additional information regarding period reliabilities that is not actually used in determining the firm yield.

8. Short-Term Conditional Reliability Modeling

The simulation discussed in the preceding sections of this chapter is based a single 1940-2012 hydrologic period-of-analysis with the six reservoirs full to capacity at the beginning of the simulation corresponding to the hydrology at the beginning of January 1940. The simulation begins during a wet period with the reservoirs full to capacity. This strategy works well for long-term planning studies and administration of the water rights permit system. However, other applications are concerned with the next one to several months or perhaps a year, with short-term future supply capabilities being highly dependent upon current reservoir storage levels.

Conditional reliability modeling (CRM) consists of developing short-term reliability and frequency metrics conditioned on preceding reservoir storage. The terms "*conditional*" and "*short-term*" are used interchangeably. CRM is based on dividing a long hydrologic period-of-analysis into many shorter simulation sequences. The simulation is repeated for each hydrologic sequence with the same initial storage condition. Water supply and hydropower reliability indices and flow and storage frequency relationships are developed from the simulation results. The WRAP programs SIM or SIMD perform the multiple short-term simulations with the specified starting reservoir storage contents. Routines in the program TABLES read the simulation results and perform reliability and frequency analyses.

CRM provides estimates of the likelihood of meeting water supply diversion, environmental instream flow, hydroelectric power, and reservoir storage requirements during time periods of one month to several months to a year or perhaps longer into the future, given current reservoir storage contents. Short-term frequencies and reliabilities are conditioned upon preceding reservoir storage levels. Applications include developing reservoir operating policies and drought management plans, operational planning studies, and decision-support during drought.

CRM uses the same input datasets as conventional WRAP applications. However, the naturalized stream flows and net evaporation depths are divided into many short sequences. For example, a 1940-2012 hydrologic period-of-analysis may be divided into 73 annual sequences. Simulations are repeated with each sequence, starting with the same initial storage conditions. Water supply reliability and storage frequency relationships are developed from the simulation results. Various options are provided for dividing the hydrologic period-of-analysis into multiple shorter sequences, assigning probabilities to the sequences, and evaluating simulation results.

The previously described frequency and reliability metrics are applied similarly with short-term CRM and conventionally long-term simulation applications. Developing storage frequency relationships is a particularly useful CRM application [29].

The example dataset is switched to CRM mode, with selected storage frequency results presented in Tables 10 and 11. The following parameters are specified in the SIM and TABLES input dataset. The simulation begins in April; time zero is the beginning of April. The storage contents of each of the six reservoirs at the beginning of April is 50 percent of the storage capacity. The beginning-of-simulation (beginning April) storage contents of the six reservoirs representing preceding or current conditions are set at 50 percent of storage capacity for each of the 73 simulations. The resulting storage frequency relationships for the end of September (six months later) are presented in Table 10. Storage frequency metrics for the end of March (12 months after beginning of April) are presented in Table 11. Table 10 is based on 73 April-September simulations performed in a single execution of SIM. Table 11 is based on 72 April-March simulations performed in a single execution of SIM. Storage contents are expressed as a percentage of storage capacity in Tables 10 and 11.

Interpretation of the storage frequency tables is illustrated as follows. Referring to Table 10, with the SIM simulations beginning in April with the initial storage contents of the six reservoirs set at 50 percent of capacity, the storage contents of Belton Reservoir equals or exceeds 15.69 percent of its 457,100 acre-feet capacity at the end of September during 90 percent of the 73 hydrologic simulation sequences. Thus, under the conditions reflected in the model, there is an estimated probability of 90 percent that the storage contents at the end of September is at least 15.69 percent of capacity. The probability that the storage volume in Belton Reservoir is at least 84.57 percent of capacity at the end of September is an estimated 20%.

Table 10. Storage Contents 6 Months Later Given 50% Full Beginning April

	Reservoir Storage Contents as a Percentage of Capacity						
	Belton	George	Granger	PK	Whitney	Waco	Total
Mean	56.20	51.21	53.66	59.19	71.87	72.33	63.52
100%	0.00	0.00	0.00	0.00	43.33	24.21	16.32
99%	0.00	0.00	0.00	0.00	45.63	30.91	18.49
98%	0.00	0.00	0.00	0.00	47.12	35.78	22.48
95%	4.01	0.00	0.00	9.03	50.28	41.44	28.97
90%	15.69	0.00	0.00	23.10	52.68	45.68	36.02
80%	35.50	15.76	12.07	37.74	57.17	49.93	45.47
70%	44.65	33.37	33.70	46.94	59.09	54.22	50.49
60%	48.12	44.85	45.39	49.89	60.44	61.63	54.83
50%	52.04	49.66	53.15	57.04	63.68	69.59	61.65
40%	62.09	59.55	63.17	64.67	70.64	84.31	68.09
30%	73.84	72.53	81.68	76.17	85.65	92.56	78.42
20%	84.57	89.39	98.50	87.52	96.18	100.00	85.10
10%	100.00	100.00	100.00	100.00	100.00	100.00	94.50

Table 11. Storage Contents 12 Months Later Given 50% Full Beginning April

	Storage Contents as a Percentage of Capacity						
	Belton	George	Granger	PK	Whitney	Waco	Total
mean	53.81	47.93	54.68	58.24	75.83	76.42	64.33
100%	0.00	0.00	0.00	0.00	43.33	18.03	16.32
99%	0.00	0.00	0.00	0.00	47.02	25.25	20.97
98%	0.00	0.00	0.00	0.00	49.60	29.64	22.63
95%	0.00	0.00	0.00	0.00	52.61	38.65	24.11
90%	2.87	0.00	0.00	11.41	56.81	44.60	32.19
80%	22.32	2.30	9.11	32.45	59.66	50.16	42.88
70%	37.51	19.71	27.27	46.52	60.44	58.53	50.49
60%	46.14	33.89	43.73	51.82	63.69	69.02	56.93
50%	52.49	46.81	54.43	58.31	71.15	84.62	65.34
40%	64.83	59.23	71.24	68.58	82.20	92.94	72.92
30%	75.20	73.35	92.32	78.52	91.35	99.30	79.98
20%	88.31	92.31	100.00	88.02	99.00	100.00	87.43
10%	100.00	100.00	100.00	100.00	100.00	100.00	95.17

The last column of Tables 10 and 11 refer to the total storage contents of all six reservoirs. Table 11 indicates that with the reservoirs 50 percent full at the beginning of April, the estimated total storage volumes 12 months later associated with exceedance probabilities of 90%, 50%, and 10% are estimated to be 32.19, 65.34, 95.17 percent of the total storage capacity of 1,949,640 acre-feet. There is a 16.32 percent chance that all six reservoirs will be completely empty 12 months later if the storage contents at the beginning of April is 50 percent of capacity.

WRAP provides flexible array of options for performing CRM. Two alternative general strategies called the equal weight method and probability array method may be applied [13, 29]. The format of the analysis results is identical with either the equal-weight or probability array approaches, but the computations are performed differently, thus yielding different numerical values. Tables 10 and 11 are based on analyses performed with the equal-weight strategy.

The equal-weight and probability array methods represent two alternative strategies for assigning probabilities to each of the short-term hydrologic simulation sequences. The equal-weight method treats each simulation sequence as one possibility out of the total number of simulation sequences. Each of the 73 hydrologic sequences are weighted the same in the analyses resulting in the frequency storage relationships of Table 10.

The probability array method is based on considering preceding reservoir storage contents in assigning probabilities to each of the hydrologic simulation sequences [13, 29]. The individual sequences are assigned different probabilities. A storage-flow-frequency relationship is developed based on applying regression to the results of a long-term SIM or SIMD simulation. The storage-flow-frequency relationship is applied in assigning probabilities to the CRM hydrologic simulation sequences. The computations are automated within TABLES.

7. Diversion Shortage Summary Metrics

The program TABLES includes an option to create an additional table with metrics summarizing failures to supply diversion requirements along with the standard routinely applied reliability table. These metrics like the others presented in this chapter have also been adopted in other models and studies reported in the literature. However, unlike all of the other metrics presented in this chapter, the following diversion shortage metrics are seldom adopted in Texas WAM system applications.

A shortage volume in a particular month of the simulation is the water supply diversion target less the simulated actual diversion as constrained by water availability. Shortages represent failures to fully meet water supply diversion requirements. A shortage period is defined as one or more consecutive months of a simulation during which a failure to meet the full diversion target occurs in each of the months. The shortage metrics are defined as follows, with the definitions varying a little between short-term CRM and long-term simulation modes.

- Maximum Shortage – The maximum shortage in any month during the entire simulation.
- Vulnerability – The average maximum shortage that occurred during each CRM simulation sequence in which a shortage occurred. In the case of a conventional long-term simulation, sequence length is defined on a January–December yearly basis.
- Resiliency – The inverse of the mean of the average length of the shortage periods that occurred during each sequence, counting only those sequences for which shortages occurred.
- Average Severity – the average sum of consecutive shortages that occurred during each sequence counting only those sequences during which shortages occurred.
- Shortage Index:

$$\text{Shortage Index} = \frac{100}{\text{months}} \sum \left(\left(\frac{\text{annual or sequence shortage}}{\text{annual or sequence target}} \right)^2 \right)$$

- Average Number of Failures per Sequence – average number of months of consecutive shortage that occurred during each simulation sequence of a CRM analysis including those sequences during which no shortage occurred. In the case of a long-term simulation, sequence length is defined on a yearly basis from January to December.
- Maximum Number of Consecutive Shortages – maximum number of consecutive months with shortages that occurred during the simulation.

The following example illustrates the above definitions. The diversion shortages occurring during three 12-month CRM simulation sequences are tabulated below and the metrics defined above are computed assuming these are the only sequences in the simulation. Sequence 1 has two short periods (January–March and December). Sequence 2 also has two shortage periods (January–March and June–October). The third sequence has only one shortage period.

Table 12. Diversion Targets and Shortages

Seq	Jan	Feb	Mar	Apr	May	Jun	Jul	Aug	Sep	Oct	Nov	Dec
					Target (acre-feet)							
1	2,733	2,649	2,775	2,901	3,447	4,414	4,667	4,457	4,204	3,742	3,111	2,901
2	2,733	2,649	2,775	2,901	3,447	4,414	4,667	4,457	4,204	3,742	3,111	2,901
3	2,733	2,649	2,775	2,901	3,447	4,414	4,667	4,457	4,204	3,742	3,111	2,901
					Shortage (acre-feet)							
1	2,250	991	1,070	0	0	0	0	0	0	0	0	2,372
2	2,505	2,404	2,775	0	0	2,478	4,667	4,457	4,204	3,742	0	0
3	2,733	1,641	0	0	0	0	0	0	0	0	0	0

maximum shortage = 4,667 acre-feet

$$\text{vulnerability} = \frac{2{,}372 + 4{,}667 + 2{,}733}{3} = 3{,}257 \text{ acre-feet}$$

$$\text{resiliency} = \frac{1}{\frac{3+1}{2} + \frac{3+5}{2} + \frac{2}{1}} = 0.375 \text{ months}$$

average severity =

$$\frac{\frac{2{,}250+991+1{,}070+2{,}372}{2} + \frac{2{,}505+2{,}404+2{,}775+2{,}478+4{,}667+4{,}457+4{,}204+3{,}742}{2} + \frac{2{,}733+1{,}641}{1}}{3}$$

$= 7{,}110.5$ acre-feet

$$\text{shortage index} = \frac{100}{36 \text{ months}} \sum \left(\left(\frac{\text{annual or sequence shortage}}{\text{annual or sequence target}} \right)^2 \right) = 15.22 \text{ months}^{-1}$$

$$\text{average number of failures per sequence} = \frac{2 + 2 + 1}{3} = 1.67$$

maximum number of consecutive months with shortages = 5 months

CONCLUSION

Effective water resources development, allocation, management, regulation, and use require comprehensive detailed capabilities for assessing water availability and reliability. Applications of the modeling and analysis methods presented in this chapter include local, regional, and statewide planning studies, administration of water allocation systems, project feasibility studies, operational planning studies, and drought management decision-support.

The general modeling strategy is based on simulations that combine historical natural river basin hydrology with a specified present or future scenario of water resources development, management, allocation, and use. Simulation results are summarized from the perspectives of likelihood, frequency, probability, percent-of-time, or reliability of meeting municipal, industrial, and agricultural water supply, environmental flow, hydroelectric energy, and reservoir storage requirements. The frequency and reliability analysis methods outlined in this chapter produce metrics for assessing water supply and management capabilities. The metrics include exceedance frequency, volume reliability, period reliability, firm yield, and diversion shortage indices.

The Water Rights Analysis Package (WRAP) is a flexible modeling system that can be applied in various ways depending on the particular water management situation. The river/reservoir system modeling strategy and frequency and reliability metrics implemented in WRAP and outlined in this chapter are applicable to other modeling systems as well.

The public domain WRAP software and documentation are readily available for application to river/reservoir systems located anywhere in the world. For river basins outside of Texas, model users must develop the input datasets required for their particular applications.

The modeling and analysis capabilities described in this chapter are routinely applied by the water management community in Texas in conjunction with the Texas WAM system. The WAM system combines the generalized WRAP with input datasets for all of the river basins of the state. The Texas Water Development Board and 16 regional planning groups apply the WRAP/WAM system in a structured legislatively mandated water planning process in which 16 regional plans and a statewide plan were created and continue to be updated in five-year cycles. The Texas Commission on Water Quality requires that water right permit applicants or their consultants apply the modeling system to assess reliabilities of proposed actions and impacts of the proposed actions on reliabilities of all other water users. TCEQ staff apply the modeling system in evaluating permit applications. River authorities, other water management agencies, consulting firms, and university researchers apply the WRAP/WAM system in operational planning studies, drought management, research studies, and other endeavors.

Development of WRAP at Texas A&M University dates back to the late 1980's. The original implementation of the WAM System by the TCEQ, its partner agencies, and contractors during 1997-2003 consisted of developing input datasets and modeling a specified set of water management and use scenarios for each of the river basins. The WRAP programs have been greatly expanded and improved in response to challenges encountered and experience gained in modeling the diverse individual river basins of the state. The WAM System datasets are continually updated by the TCEQ to reflect changes in water right permits and water management practices and refinements in modeling capabilities. WRAP/WAM development efforts at Texas A&M University during 2011-2014 focus on supplementing the monthly modeling system with a daily version, improving capabilities for modeling environmental flow requirements and their impacts on other water users, simulating flood control operations, and developing efficient methods for updating hydrology. WRAP modeling tools can be adopted and the Texas experience built upon to address water management issues throughout the world.

References

[1] Maass, A.; Hufschmidt, M.M.; Dorfman, R.; Thomas, H.A.; Marglin, S.A; Fair, G.M. Design of Water Resource Systems; Harvard University Press: Boston, MA, 1966.

[2] Votruba, L; Broza, V. Water Management in Reservoirs; Developments in Water Science, Vol. 33, *Elsevier Science:* Philadelphia, PA, 1989.

[3] Wurbs, R.A. Modeling and Analysis of Reservoir System Operations; Prentice-Hall: Upper Saddle River, NJ, 1996.
[4] ReVelle, C.S; Optimizing Reservoir Resources; John Wiley & Sons: New York, NY, 1999.
[5] Nagy, I.V.; Asante-Duah, K.; Zsuffa, I. Hydrologic Dimensioning and Operation of Reservoirs: Practical Design Concepts and Principles; *Water Science and Technology Library* Vol. 39, Kluwer Publishers: London, UK, 2002.
[6] Karamouz, M.; Szidarovszky, F.; Zahraie, B. Water *Resources Systems Analysis;* Lewis Publishers: Boca Raton, FL, 2003.
[7] Jain, S.K.; Singh, V.P. Water Resources Systems Planning and Management; Developments in Water Science 51, *Elsevier:* Philadelphia, PA, 2003.
[8] Simonovic, S.P.; Managing Water Resources: Methods and Tools for a Systems Approach; *United Nations Educational, Scientific and Cultural Organization*, Paris, France, 2009.
[9] Bhave, P. R. *Water Resources Systems*; Narosa Publishing: Oxford, UK, 2011.
[10] Labadie, J.W. Optimal operation of multireservoir systems: state-of-the-art review; *Journal of Water Resources Planning and Management*, ASCE, 130(2), 93-111, 2004.
[11] Wurbs, R.A. Comparative Evaluation of Generalized River/Reservoir System Models; TR-282, 203 pages, *Texas Water Resources Institute*: College Station, TX, 2005.
[12] Wurbs, R.A. Chapter 1 Generalized Models of River System Development and Management; *Current Issues in Water Management* (U. Uhlig, Editor), pp 1-22, InTech, Rijeka, Croatia, 2011.
[13] Wurbs, R.A. Water Rights Analysis Package Modeling System Reference Manual; TR-255, 10th Edition, 384 pages, *Texas Water Resources Institute*: College Station, TX, 2013.
[14] Wurbs, R.A. Water Rights Analysis Package Modeling System Users Manual; TR-256, 10th Edition, 185 pages, Texas Water Resources Institute: College Station, TX, 2013.
[15] Wurbs, R.A. Fundamentals of Water Availability Modeling with WRAP; TR-283, 7th Edition, 120 pages, *Texas Water Resources Institute*: College Station, TX, 2013.
[16] Wurbs R.A. WRAP River System Hydrology; TR-431, 2nd Edition, 195 pages, *Texas Water Resources Institute*: College Station, TX.

[17] Wurbs R.A. Salinity Simulation with WRAP; TR-317, 87 pages, *Texas Water Resources Institute*: College Station, TX, 2009.
[18] Wurbs R.A, Hoffpauir R.J. Water Rights Analysis Package Daily Modeling System; TR-430, 2nd Edition, 285 pages, *Texas Water Resources Institute*: College Station, TX, 2013.
[19] Wurbs, R.A.; Hoffpauir, R.J. Water Rights Analysis Package Modeling System Programming Manual; TR-388, 3rd Edition, 196 pages, *Texas Water Resources Institute*: College Station, TX, 2013.
[20] HEC-DSSVue HEC Data Storage System Visual Utility Engine, User's Manual, Version 2.0; CPD-78, *Hydrologic Engineering Center*, U.S. Army Corps of Engineers, Davis, CA, 2009.
[21] WRAP Display Tool, Users Manual and Operating Instructions*; Center for Research in Water Resources, University of Texas at Austin*; Texas Commission on Environmental Quality: Austin, TX, 2012.
[22] Wurbs, R.A.; Hoffpauir, R.J. Environmental Flows in Water Availability Modeling; TR 440, 285 pages, *Texas Water Resources Institute*: College Station, TX, 2013.
[23] Water for Texas 2012 State Water Plan; *Texas Water Development Board*: Austin, TX, 2012.
[24] Wurbs, R.A. Texas water availability modeling system, *Journal of Water Resources Planning and Management,* ASCE, 131(4):270-279; 2005.
[25] Wurbs, R.A.; Hoffpauir, R.J.; Schnier, S.T. Application of Expanded WRAP Modeling Capabilities to the Brazos WAM; TR-389, 2nd Edition, 356 pages, *Texas Water Resources Institute*: College Station, TX, 2012.
[26] Wurbs, R.A.; James, W.P. *Water Resources Engineering*, Prentice Hall, Upper Saddle River, NJ, 2002.
[27] HEC-SSP Statistical Software Package, User's Manual, Version 2.0; *Hydrologic Engineering Center*, U.S. Army Corps of Engineers, Davis, CA; 2010.
[28] Wurbs, R.A.; Lee, C.H. Salinity in water availability modeling; *Journal of Hydrology, Elsevier Science*, 407(2), 451-459, 2011.
[29] Wurbs, R.A.; Schnier, S.T.; Olmos, H.E. Short-term reservoir storage frequency relationships; *Journal of Water Resources Planning and Management*, ASCE, 138(6), 597-605, 2012.

In: Drinking Water and Water Management ISBN: 978-1-63117-523-7
Editor: Mason Giannino

Chapter 2

EFFICIENT ADVANCED OXIDATION TECHNOLOGIES AGAINST CHEMICAL THREAT IN DRINKING WATER DISTRIBUTION SYSTEMS

Mónica S. F. Santos, Cátia Oliveira, Arminda Alves and Luis M. Madeira*

LEPABE – Laboratory for Process, Environmental, Biotechnology and Energy Engineering; Faculty of Engineering, University of Porto, Porto, Portugal

ABSTRACT

Vulnerability of drinking water distribution systems (DWDS) is one of the main issues of concern to regulatory agencies and water utilities. Indeed, the threat of a deliberate (or even accidental) contamination of water supplies is real and is permanently under consideration. Although measures against such contamination events have been intensified in the last few years, major knowledge gaps still exist in what concerns the feasibility of the detection and remediation methods, especially those related with contamination by chemicals. This has been recently addressed in a European Project named SecurEau (www.secureau.eu)

* Corresponding author. Tel.: +351-225081519; Fax: +351-225081449;
E-mail: mmadeira@fe.up.pt.

whose main objective was to launch an appropriate response for rapidly restoring the use of a DWDS after a contamination event. This chapter intends to present the developments made concerning the remediation methods, proposing several decontamination procedures, based on advanced oxidation processes (AOPs), for the treatment of waters/solids contaminated with key chemicals. Based on several criteria, three model compounds were selected: an organophosphate (chlorfenvinphos - CFVP), an herbicide (paraquat - PQ) and a brominated flame retardant (2,2´,4,4´,6-pentabromodiphenyl ether (BDE-100)).

The treatment of waters contaminated with PQ was studied by both Fenton and photo-Fenton process under off-line conditions (batch reactor) and by Fenton´s process under on-line conditions (i.e., in-situ of a pilot loop reactor, simulating a DWDS). Out of the pipelines, the photo-Fenton approach has great advantages over Fenton´s process since it allows a significant increase on the PQ degradation rate and, particularly, on the mineralization degree (96% after 1 h of reaction). Concerning the in-situ treatment, the Fenton's reagent proved to be effective for the treatment of PQ-contaminated waters, and it was shown that some deposits and pipe materials could be used as catalysts, avoiding the addition of other chemicals inside the network.

Water and deposits contaminated with CFVP can also be treated in-situ by Fenton's reagent making use of the active iron available either on pipe materials, on deposits attached to them or on loose deposits as catalysts of the process.

The direct photolysis methodology was successfully applied to the off-line remediation of BDE-100 contaminated waters. Batch experiments with pure water revealed a high BDE-100 degradation profile for the first five minutes of exposition to UV-light (1.6×10^{-6} Einstein/s).

Therefore, it has been proved that several solutions exist using AOPs for the effective treatment of waters and deposits contaminated with various chemicals, with different properties; this might be an important step forward towards chemical threats in drinking water distribution systems.

1. INTRODUCTION

Although our planet is called "blue planet" (once 70% of it is covered with water), only 2% is freshwater for human use, because the major part corresponds to oceans, where the water is too salty to drink and inadequate for many applications. Of the freshwater available on Earth, about 99% is in ice and groundwater (water underground in aquifers). Although groundwater resources exceed salt-free surface water on earth, typically only individual households and small towns may be served sufficiently by groundwater from wells or springs. Large cities tend to use surface water and centralized water treatment and distributions systems. The quantity of surface water available to drink is dramatically affected by the seasons, weather patterns and long-term shifts in climate. Additionally, surface water is more likely to suffer anthropogenic contamination than groundwater due to its proximity to human activities. Surface, and in some cases groundwater, could be polluted by exposition to acid rain, storm water runoff, sewage overflow, agricultural runoff and industrial waste. Microbial pollution of drinking water has been considered the most important threat for human health [1], but there are well established and robust methods for protecting drinking water from that [2]. On the other hand, while the consequences of microbial contamination are quickly and fairly easily identified, chemical contamination is usually much more insidious, with effects not seen until the source has been in use, usually for a considerable large period.

Water plays an important role in our life, being imperative to human and ecosystem health, and is the basis for an economic welfare. Due to the extreme importance of this source, water and particularly drinking water has often been a political or military target [3]. For that also contributes the several points of vulnerability of a water distribution system: the catchment, the raw water transport system, the treatment facilities, the water reservoirs and the distribution systems. In this perspective, Peter H. Gleick (2006) reports a lot of situations where water or systems were used as delivery vehicles to cause violence to a human population [4]. A great concern about this issue has been generated and a lot of measures have been taken to limit the impact that these kinds of attacks could represent for human life. Protection and surveillance of water infrastructures, the detection of water quality deterioration in drinking water distribution systems and operational procedures for decontamination of water infrastructures are needed to prevent or to restore quickly the functionality of the distribution system after a deliberate (or accidental) contamination. In this context, and with some experience gained in the past with other related projects, SecurEau, an European project (FP7-SEC-2007-1) entitled "Security and decontamination of drinking water distribution systems following a deliberate contamination", was launched [5]. The intent of SecurEau was to serve as a demonstration project for designing and implementing an effective and timely response on CBRN (chemical, biological and radionuclide) agents. Part of the work done during the four years of the project duration, and now presented in this book chapter, had the intention to identify alternative remediation methods to quickly eliminate the contamination, especially that originated by emergent chemicals.

The work present herein was focused in three model chemicals (an herbicide, an organophosphate pesticide and a flame retardant) selected by SecurEau team. Paraquat (herbicide), chlorfenvinphos (organophosphate pesticide) and BDE-100 (flame retardant) were chosen to be target compounds due to their different physical-chemical characteristics, which determine the final effects in consumers, as well as in sorption, dispersion and accumulation patterns in case of a deliberate contamination of drinking water distribution systems. Paraquat is included in the "old" toxicants that may be easily acquired and manipulated, despite of the effective control measures. The organophosphate pesticides are considered as nerve agents due to their cholinergic action, producing acetylcholinesterase inhibition, shown by the clinical effects called SLUDGE – Salivation, Lacrimation, Urination, Defecation, Gastrointestinal distress and Emesis. It is believed that the nerve agents, already available but not used during the World War II, are still the most concerning chemical weapons for terrorism. On the other hand, they are still widely used in most of underdeveloped countries and therefore may be easily and uncontrollably sold in commercial stores, thus prone to be used in such attacks. Finally, BDE-100 was chosen to represent the "new" chemical compounds with lethal effects even at extremely low doses.

A state of the art review concerning the existing degradation methodologies for the treatment of waters contaminated with these target chemicals was performed and is presented below.

1.1. Paraquat

Paraquat (PQ) was first synthesized in 1882 [6]. Since 1933, it has been used as a redox indicator (under the name methyl viologen) in chemical laboratories [6]. Its herbicidal properties were discovered in 1957 and one year later PQ was introduced by Imperial Chemical Industries [6]. In 1962, PQ began to be marketed as a highly effective contact herbicide [6]. Nowadays, this bipyridylium compound is widely used as herbicide around the world (nearly 90 countries), despite of being prohibited in Europe [7]. Its popularity is related to its quick and non-selective action to kill green plant tissue upon contact. Some studies proved that PQ is one of the few herbicides capable of controlling the growth of weeds that became resistant as a result of over-use of non-selective glyphosate herbicides [8, 9].

Table 1. Physical and chemical properties of paraquat[a]

Chemical structure Common Name CAS n.º Molecular formula Molecular weight	Solubility in water at 25 °C (g/L)	Melting point (°C)	Boiling point (°C)	Vapor pressure (mmHg)	Henry's constant (atm. m^3/mol)	Log K_{ow}
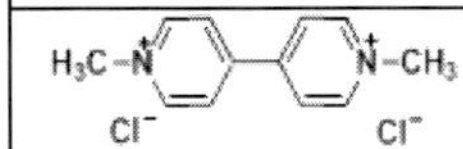 Paraquat 1910-42-5 $C_{12}H_{14}N_2Cl_2$ 257.18 g/moL	620	108.59	351.92	1.01×10^{-7}	3.22×10^{-13}	-4.5

[a] Values were extracted from EPI Suite TM, copyright 2000-2012 Environmental Protection Agency, United States.

PQ is denoted as a class I toxicological agent, being considered a possible carcinogenic and mutagenic molecule [10-12]. Its toxicity is associated with the reduction of the bivalent cation (PQ^{2+}) to a free radical ($PQ^{\bullet+}$), which reacts with oxygen to form a superoxide radical ($O_2^{-\bullet}$), hydrogen peroxide (H_2O_2) and hydroxyl radical ($OH^{\bullet}$) [12]. Since the produced radicals are unstable and reactive, they cause lesions in the cellular membrane, protein and DNA [13]. PQ is often encountered in cases of poisoning [14-16]. Chronic and acute exposition effects caused by casual ingestion or poisoning with PQ are well known and have been documented in the literature [17-19]. Some of these studies report respiratory distress [16, 20], neurodegenerative disorder of the nervous system (such as Parkinson's disease) [21, 22] and effects on the kidneys [20].

The uncontrolled and abusive use of PQ has generated a great concern related to the potential risk that it represents to humans, animals and the environment [23]. For such feelings have contributed the long residence time of PQ in the environment (it is a non-biodegradable and highly persistent molecule), its elevated toxicity, even at low doses, its large availability and its relatively low cost [10, 11, 24].

The main physical-chemical characteristics of PQ are indicated in Table 1, where it can be seen that this chemical has extremely high water solubility. This property, together with those presented before, such as availability, low cost and high toxicity, support the idea that it could be a strong candidate to be used in a deliberate contamination; this is the reason why it was selected as model compound in SecurEau project. Additionally, this polar herbicide is positively charged, non-volatile, thermally stable and stable in acidic conditions [25].

In the European Union drinking water legislation there are no established values specifically for PQ, but a maximum individual pesticide concentration in drinking water was set at 0.1 μg/L [26].

1.1.1. Degradation Technologies for Paraquat in Waters

To reduce the PQ (and other pesticides) impact, powerful and practical treatment processes for its (their) degradation in water matrices are needed. The traditional chemical and biological treatments are not sufficiently effective for most pesticides [27]. On the other hand, in some physical processes there is a simple transfer of the pollutants from one phase to another, not being destroyed. The advanced oxidation processes (AOPs) are clean technologies based on the generation of extremely reactive and non-selective hydroxyl radicals, with very high oxidative power. Due to their powerful capability to oxidize numerous organic compounds into CO_2 and H_2O, AOPs have been selected for various applications. Among them, the Fenton's reaction is a promising process, which appeared in 1894 when Fenton strongly improved tartaric acid oxidation by the use of ferrous ion and hydrogen peroxide [28]. Fenton's reaction is a non-expensive and environmental friendly oxidation method, widely studied for wastewater treatment. The process has a complex mechanism, which can even though be simplified by the following equations [29, 30]:

$$H_2O_2 + Fe^{2+} \rightarrow Fe^{3+} + OH^- + HO^\bullet, \quad k_1 = 76 \ M^{-1}s^{-1} \tag{1}$$

$$Org. + HO^\bullet \rightarrow Org.^\bullet \rightarrow ... \rightarrow CO_2 + H_2O, \ k_2 \approx 10^7 - 10^{10} M^{-1}s^{-1} \tag{2}$$

$$H_2O_2 + OH^\bullet \rightarrow H_2O + HO_2^\bullet, \quad k_3 = (1.2 - 4.5) \times 10^7 M^{-1}s^{-1} \tag{3}$$

$$Fe^{2+} + HO^\bullet \rightarrow Fe^{3+} + OH^-, \quad k_4 = 4.3 \times 10^8 M^{-1}s^{-1} \tag{4}$$

$$HO^\bullet + HO^\bullet \rightarrow H_2O_2, \quad k_5 = 5.3 \times 10^9 M^{-1}s^{-1} \tag{5}$$

$$Fe^{3+} + H_2O_2 \leftrightarrow Fe^{III}(HO_2)^{2+} + H^+, \quad k_6 = 3.1 \times 10^{-3} \tag{6}$$

$$Fe^{III}(HO_2)^{2+} \rightarrow Fe^{2+} + HO_2^\bullet, \quad k_7 = 2.7 \times 10^{-3} s^{-1} \tag{7}$$

Briefly, the reaction between iron and hydrogen peroxide produces hydroxyl radicals with high oxidative power (eq. 1), which attack the organic matter present in the water (eq. 2). Unfortunately, some parallel reactions occur (eqs. 3-5), and so the hydroxyl radicals are not only consumed to degrade the organic matter but also to produce other radicals, with less oxidative power, or other species (so-called scavenging effect of $HO^\bullet$). Besides, this leads to the undesired consumption of hydrogen peroxide (eq. 3). On the other hand, eqs. 6 and 7 indicate the regeneration of Fe^{2+} by the reaction between H_2O_2 and Fe^{3+} (Fenton-like process); this way ferrous ion is restored, acting as catalyst in the overall process.

Table 2. Studies reported in the literature about paraquat degradation in water by AOPs*.

Treatment	Operating conditions	Analytical method for PQ monitoring	Results and Comments	Ref.
UV light over TiO_2-coated polythene or polypropylene films	$[PQ]_0 = 50$ mg/L T = 26 °C 15 mL O_2/min 8.8×10^{-5} g TiO_2/cm^2 for polythene and 3.1×10^{-5} g TiO_2/cm^2 for polypropylene	UV spectrophotometer	Complete degradation to CO_2, NH_3, HCl and small quantities of NO_2^-/NO_3^- within ca. 6 h.	[31]
UV light over commercial TiO_2	$[PQ]_0 = 10$, 20 and 40 mg/L UV (365 nm) 50 mL O_2/min 200 mg TiO_2/L pH = 4, 6.6, 7 and 9	LC	Paraquat is slowly degraded by direct photolysis in the presence of dissolved oxygen ($[PQ]_0 = 20$ and 40 mg/L; pH_0 6.6). Direct photolysis with UV light destroyed 60% of paraquat in less than three hours of reaction ($[PQ]_0 = 10$ mg/L; pH_0 6.6). In the presence of TiO_2, all species were consumed after three hours of reaction at high pH values.	[32]
UV light over TiO_2	$[PQ]_0 = 1000$ mg/L 0.5 L air/min TiO_2 film (three-times coating) attained from the hydrothermal method UV-light intensity: 36 W/m^2	Potassium ferric-oxalate actinometrical method	About 100% of paraquat was degraded after 15 h of reaction.	[33]

* Reprinted from Chemical Engineering Journal, volume 175, 279-290, Mónica S. F. Santos, A. Alves, Luís M. Madeira, Paraquat removal from water by oxidation with Fenton's reagent. Copyright (2011), with permission from Elsevier.

Treatment	Operating conditions	Analytical method for PQ monitoring	Results and Comments	Ref.
UV light and/ or air-sparging over TiO_2	$[PQ]_0 = 100$ mg/L 1 L air/min 0.1 g TiO_2/L UV-light intensity: 4, 8, 12, 24 and 36 W/m^2	UV – spectrophotometer	Paraquat removal rate was 0.54 mg/L/h with only air-sparging. In the presence of UV-light, the removal rate with air-sparging was 50% higher in 24 h than that without it. In the presence of TiO_2, the removal rates in 40 h were 1.4, 1.6 and 2.2 mg/L/h at the UV light intensities of 4, 8 and 12 W/m^2, respectively. In the presence of TiO_2 with air-sparging, the required times for 90% removal of PQ were 18, 12 and 3 h at the UV light intensities of 12, 24 and 36 W/m^2, respectively.	[41]
UV light over TiO_2 (commercial and "home prepared")	$[PQ]_0 = 7.5$ mg/L 200 mg TiO_2 /L pH = 4, 7 or 9	UV-visible spectrophotometer	Under alkaline medium, paraquat degradation is almost complete after 30 min of reaction. Five compounds structures were proposed for intermediate degradation products of paraquat oxidation	[34]
UV light over commercial TiO_2	$[PQ]_0 = 20$ mg/L 0.04 and 0.4 g TiO_2/L UV-light intensity: 140 W/m^2	UV spectroscopy	Near complete mineralization of paraquat was achieved after ca. 3 h of irradiation by using 0.4 g/L of catalyst amount at neutral pH (ca. 5.8).	[35]
Xenon lamp over $1C_{60}$/ 1 V-TiO_2 (composition in wt%)	$[PQ]_0 = 15 – 50$ mg/L 1 g catalyst/L pH=6.5	UV-visible spectrophotometer	70% degradation within 4 h.	[36]
UV light in the presence of O_2 or O_3	$[PQ]_0 = 1500$ mg/L 32 g O_3 /h 66 low-pressure mercury vapor lamps (254 nm)	GC	Significant oxidation of PQ did not occur in the presence of oxygen during 7 h. When O_3 was fed 32.8% of PQ was destroyed during 7 h. 4-carboxy-1-methylpyridinium ion, monoquat, monopyridone, 4,4'-bipyridyl, 4-picolinic acid, oxalate, malate, succinate and N-formylglycine were proposed for intermediate compounds of PQ oxidation.	[37]

Table 2. (Continued)

Treatment	Operating conditions	Analytical method for PQ monitoring	Results and Comments	Ref.
O_3	$[PQ]_0 = 231 – 2057$ mg/L 38.6 L O_3/h pH = 4.2 to 8.0	LC	Paraquat was degraded after 120 min of reaction at pH 8.0. N-methylisonicotinic acid, N-formyloxamic acid, ammonia, oxamic acid and nitrate were proposed for intermediate compounds of PQ oxidation.	[38]
Heterogeneous Fenton (AC-Fe as catalyst)	$[PQ]_0 = 20$ mg/L 1 g catalyst/L T=70 °C $[H_2O_2]_0 = 12.5$ mM	LC	71.4% of chemical oxygen demand abatement was obtained under the conditions indicated.	[39]
Anodic oxidation (AO); Electro-Fenton (EFe); Photoelectro-Fenton (PEF)	$[PQ]_0 = 10 – 50$ mg/L T = 25 °C pH = 3 $[Fe^{2+}]_0 = 0.1 – 2$ mM $[H_2O_2]_0 = 20$ mM Current intensity: 50, 100 and 200 mA	LC	The PEF is the most efficient treatment. Classic dark Fenton under the same conditions leads to a high removal rate at the initial stage of the experiment but poor degradation efficiency was obtained maybe due to the insufficient H_2O_2.	[40]
Homogeneous Fenton	$[PQ]_0 = 50 – 200$ mg/L T = 10 – 70 °C pH = 2 – 5 $[Fe^{2+}]_0 = 1.0\times10^{-4} – 5.0\times10^{-4}$ M $[H_2O_2]_0 = 1.6\times10^{-3} – 5.7\times10^{-2}$ M	LC	Under selected conditions (T =70.0 °C, $[Fe^{2+}] = 5.0\times10^{-4}$ M, $[H_2O_2] = 3.4\times10^{-2}$ M, and pH = 3.0) nearly all the initial paraquat dichloride was degraded after 1 h and 60% of the organic matter was mineralized after 4 h. At 30 °C, complete paraquat degradation and 40 % of mineralization were reached after 4 h of reaction.	[42]

Notes: AO – anodic oxidation; EFe – electro-Fenton; GC – gas chromatography; LC – liquid chromatography; PEF – photoelectro-Fenton; UV – ultraviolet.

Table 3. Physical and chemical characteristics of CFVP[a]

Chemical structure Common Name CAS n.º Molecular formula Molecular weight	Solubility in water at 20 °C (mg/L)	Melting point (°C)	Boiling point at 0.001 mm Hg (°C)	Vapor pressure at 25 °C (mbar)	Henry's constant at 20 °C (atm. m^3/mol)	Log K_{ow}
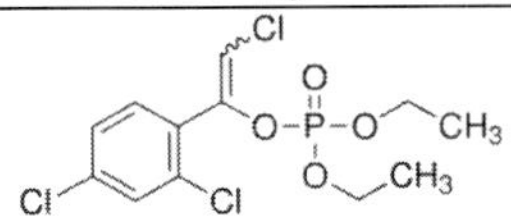 Chlorfenvinphos 470-90-6 $C_{12}H_{14}Cl_3O_4P$ 359.57 g/mol	124	-20	167-170	7.50×10^{-6}	5.17×10^{-8}	3.81

[a] Values were extracted from EPI Suite TM, copyright 2000-2012 Environmental Protection Agency, United States.

Table 4. Studies reported in the literature about chlorfenvinphos degradation in water by AOPs

Treatment	Operating conditions	Analytical method for CFVP monitoring	Results and Comments	**Ref.**
Photo-Fenton	$[CFVP]_0 = 50$ mg/L T = 25 – 35 °C $pH_0 = 2.6$ UV-light intensity: 22 W/m^2 $[H_2O_2]_0 = 1$ mM $[Fe_{3+}] = 0.36$ mM	LC-UV	Complete degradation achieved (concentration achieved below the LC system detection limit)	[47]
Ozonation	pH = 2 T = 10 – 40 °C $[CFVP]_0 = 1$ μM $[O_3]_0 = 3 \times 10^{-5} – 1.0 \times 10^{-4}$ M	LC-DAD	After 30 minutes, the conversion in CFVP attained was 76 %; this result was obtained under 40 °C and $[O_3]_0 = 5 \times 10^{-5}$ M.	[48]
UV radiation	$[CFVP]_0 = 10$ μM pH = 3 – 9 T = 20 – 40 °C λ = 254 nm	LC-DAD	A CFVP conversion of 88.5 % was obtained after 4 minutes, when CFVP was submitted to a pH medium of 7, at 40 °C.	[48]
O_3/H_2O_2	$[CFVP]_0 = 1 \times 10^{-6}$ M T = 20 °C pH = 7 $[O_3]_0 = 41.6$ μM $[H_2O_2]_0 = 20.8$ μM	LC-DAD	After 150 seconds, for the described conditions, CFVP was found in a concentration of 2×10^{-7} M.	[48]
UV/ H_2O_2	$[CFVP]_0 = 1 \times 10^{-6}$ M T = 20 °C pH = 7 λ = 254 nm $[H_2O_2]_0 = 2 \times 10^{-4}$ M	LC-DAD	Complete removal of CFVP was attained after 10 minutes of light exposure.	[48]

Treatment	Operating conditions	Analytical method for CFVP monitoring	Results and Comments	Ref.
Photo-Fenton	$[CFVP]_0 = 50$ mg/L [Fe] = 2 and 55 mg/L UV-light intensity: 30 W/m^2 pH = 2.7 – 2.8	LC-UV	Complete CFVP degradation was attained in seconds, while mineralization was achieved after 50 minutes, when a concentration in iron of 55 mg/L was applied. Nevertheless, using lower iron concentrations (2 mg/L) is enough to eliminate the pesticide from water.	[49]
TiO_2 photocatalysis	$[CFVP]_0 = 50$ mg/L $[TiO_2] = 200$ mg/L UV-light intensity: 30 W/m^2 pH = 2.7 – 2.8	LC-UV	When TiO_2 is applied, a result similar to the one obtained using 2 mg/L iron is attained (more than 80 % of mineralization after 100 minutes).	[49]
Photo-Fenton	$[CFVP]_0 = 30$ mg/L $[Fe^{2+}] = 2$ mg/L UV-light intensity: 80 mW/cm^2 pH = 2.7	LC-UV	96 % of CFVP eliminated after 0.15 h (solution of contaminants composed by CFVP and more 4 pesticides).	[50]
Photo-Fenton/O_3	T = 25 °C $[CFVP]_0 = 50$ mg/L pH = 3 $[Fe^{2+}] = 80$ mg/L $[H_2O_2]_0 = 0.26$ g/L $Q_{O3} = 1.6$ g/L	LC-UV vis	Complete degradation attained after 15 min; for the same period of time, 80 % mineralization was achieved.	[51]
TiO_2 photocatalysis/ O_3	T = 25 °C $[CFVP]_0 = 50$ mg/L pH = 3 $[TiO_2] = 0.25$ g/L	LC-UV vis	Elimination of 80 % of CFVP and 60 % mineralization after 10 minutes.	[51]

Notes: DAD – diode array detector; LC – liquid chromatography; UV – ultraviolet detector.

Few studies about PQ degradation by AOPs in water were published and are hereby compiled in Table 2. As can be seen in Table 2, most of the works done addressing PQ degradation by AOPs refer to photocatalytic processes by the action of TiO_2 [31-36]. In these cases, high PQ degradation performances were achieved for a wide range of initial parent compound concentrations (7.5 to 1000 mg/L) and experimental conditions. Nevertheless, it is important to emphasize that this kind of treatment could import excessive electricity costs, unless it makes use of solar radiation. Ozonation was also tested to degrade PQ in water [37, 38]. Relatively low PQ degradation by ozonation in the presence of light was obtained by Kearney et al. [37] (only 32.8% of PQ was destroyed after 7 h). More recently, Dhaouadi and Adhoum [39] used four different activated carbons as catalysts in PQ degradation by the heterogeneous Fenton's process. The same researchers also tested and compared four different advanced oxidation methods in the degradation of PQ: anodic oxidation (AO), electro-Fenton (EFe), photoelectro-Fenton (PEF) and classic Fenton (CFe) [40]. All but one treatment method (EFe) was optimized. The CFe treatment was not optimized as only one experiment was done for comparison purposes. The PQ concentration versus time plot indicated that the fastest initial degradation rate was attained for the CFe method, but after 20 min no further PQ degradation was observed. Dhaouadi and Adhoum [40] concluded that CFe has a significantly lower oxidative ability as compared to other systems. However, this may be due to the insufficient peroxide quantity used in the operation. Therefore, an optimized classic Fenton process could potentially lead to comparable PQ degradation efficiencies. Up to the author's knowledge, there are no other studies reported concerning PQ degradation by the homogeneous Fenton's process, which was addressed by the team and is summarized in this book chapter (cf. section 2).

1.2. Chlorfenvinphos

Chlorfenvinphos (CFVP) is an organophosphorus chemical that acts as insecticide and acaricide. CFVP has been used to control the insect plagues in agriculture. Nowadays, due to its high neurotoxicity, it was banned and is only applied to veterinary products (as a controller of parasites in animals) [43]; nevertheless, it is still used in some countries as Australia or Switzerland [44].

Chlorfenvinphos was for the first time introduced in the United States in the 60's. Its toxicity is associated to a cholinesterase enzyme inhibitor, causing breath and nerve damages, leading, in the end, to the death of insects. In humans, this compound can damage several organs (eyes, liver, kidneys and even the cardiovascular and reproductive systems) [45], producing several symptoms as the increasing salivation, changes in blood pressure and heart beating, nauseas, diarrhea, headaches, and, in very high doses, ıt can cause death. Furthermore, according to the directive 2000/60/EC [46], CFVP is one of the 33 substances considered priority because of its high toxicity and persistence in the environment. Table 3 shows the main physical-chemical properties of this compound, from which the relatively high water solubility is emphasized (although much lower than that for paraquat, as shown before).

Taking special attention to the devastating effects of this pesticide, it is necessary to develop technologies capable of eliminating it from contaminated waters and deposits.

1.2.1. Degradation Technologies for Chlorfenvinphos in Waters

CVFP can easily pass to ground water due to its large use in agriculture/veterinary acts, being gradually sorbed in the soils, contaminating waters and, consequently, pipes and deposits in the water treatment systems. Furthermore, due to its toxicity, it was considered as a model compound (in the framework of the above-mentioned SecurEau project) in case of a deliberate contamination, once the concentrations attained could be quite high (due to its relatively high water solubility) with hazardous and tremendous potential impacts.

Several studies were performed aiming the decontamination of waters contaminated with CFVP, which are summarized in Table 4. Among them, it is worth mentioning the use of ozonation, UV/Vis-based processes (like photo-Fenton or TiO_2-assisted catalysis) and hybrid ones (e.g., O_3/H_2O_2), being the final performance dependent not only on the process but also on the conditions employed. Fenton's oxidation was not reported except by the authors, as mentioned in section 3.

1.3. Polybrominated Diphenyl Ethers (PBDEs)

Accidental fires inflict a heavy toll in terms of economic loss, human suffering and, in the extreme, death. As a response to this challenge, there was a need to produce and use flame retardants (FR), as a mean to reduce the likelihood of ignition, to hinder the fire spread and to provide some extra time in the early stages of a fire when it is much easier to escape [52]. For a long time, flame retardants were associated to safe and protection and were used in a widespread number of applications such as: industries of plastics, textiles, electronic equipment and in building materials. Among the organohalogenated flame retardants, the bromine-based ones are the most effective and most widely applied [53]. Their popularity on fire prevention is related in part to the weak bond between bromine and the carbon atoms, which enables the bromine atom to interfere at a more favorable point in the combustion process [52].

Brominated flame retardants (BFRs) can be classified in reactive or additive, depending on the manner they are incorporated in the polymer. Reactive flame retardants are chemically bonded to the material while additive flame retardants are not covalently bonded and are often applied to the substrate surface as a spray in a coating formulation [54]. For that reason, additive BFRs are easily leached out from the surface of the material and thus released into the environment during their natural operational life and also during processing, recycling or combustion [52, 55]. Polybrominated diphenyl ethers (PBDEs) are an example of additive flame retardants [56]. As observed in Figure 1, they have structures similar to polychlorinated biphenyls (PCBs) with an oxygen atom between the aromatic rings [56].

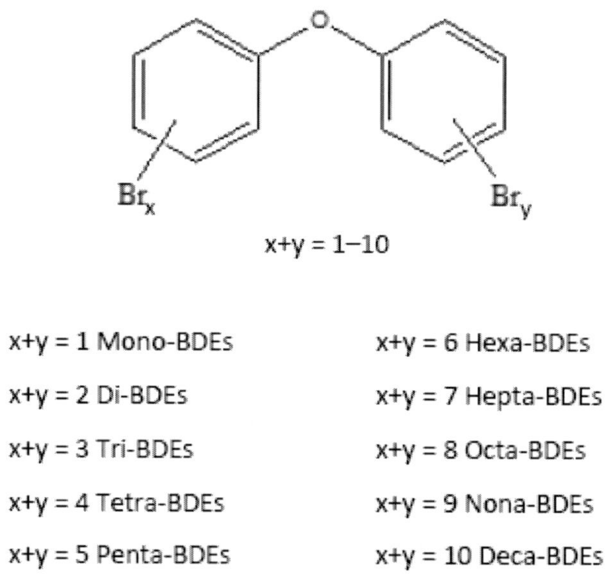

Figure 1. Chemical structure of PBDEs (adapted from [57]).

Theoretically, there are 209 PBDE congeners differing in the number and/or position of the bromine atoms in the aromatic ring. Commercial PBDEs are typically prepared by bromination of diphenyl ether in the presence of a catalyst. As a consequence, mixtures of PBDEs with various bromination degrees are produced. There are three main PBDE commercial mixtures: penta-BDE, octa-BDE and deca-BDE. Commercial penta-BDE is a mixture of tetra (24-38%), penta (50-60%) and hexa (4-8%) congeners [58]. The octa-BDE mixture is composed by hexa (10-12%), hepta (44%), octa (31-35%) and nona (10-11%) congeners [58]. Commercial deca-BDE formulation contains octa (0.8%), nona (22%) and deca (77%) congeners [59]. The main properties of technical PBDE mixtures are indicated in Table 5. The uses of penta-BDE and octa-BDE commercial mixtures were completely banned by the European Union (EU) since 2004, after a comprehensive risk assessment analysis under the Existing Substances Regulation 793/93/EEC [60]. In May 2009, at the 4th meeting of the parties of the Stockholm Convention for Persistent Organic Pollutants (POPs), the penta-BDE and octa-BDE commercial mixtures were officially classified as POPs substances and were included in the Annex A (elimination of production and use of all intentionally produced POPs) [61, 62]. In the United States, several states have adopted legislation to ban the penta-BDE and octa-BDE mixtures and, owing to growing apprehension, the main US chemical producer voluntarily agreed to stop its manufacturing [63]. Deca-BDE commercial mixture has been explored without any restriction which, in part, has contributed to the increase on the lower PBDEs environmental findings as a consequence of congeners debromination [63]. Indeed, the major PBDEs findings in environmental matrices correspond to lower brominated flame retardants, which have been considered more dangerous for humans and wildlife than the higher brominated ones due to their high efficiency to be bioaccumulated [59, 64-66]. The larger molecules, specially the fully brominated deca-BDE, are poorly absorbed due to their lower ability to cross biological membranes, being rapidly eliminated, and thus they are unlikely to be bioaccumulative [52, 56]. Due to their persistence, bioaccumulation and toxicity, ubiquity and long-range atmospheric transport, PBDEs have even been found in the remote arctic environment [67]. Estimated half-lives in the environment between 2 and 10 years were reported for these chemicals [68].

Table 5. Physical and chemical properties of technical PBDE mixtures (adapted from [69])

Property	penta-BDE	octa-BDE	deca-BDE
Color	Clear, amber to pale yellow	Off-white	Off-white
Physical state	Highly viscous liquid	Powder	Powder
Melting point	-7 to -3 °C (commercial)	85-89 °C (commercial)	290-306 °C
Boiling point	>300 °C (decomposition starts above 200 °C)	decomposes at >330 °C (commercial)	decomposes at >320 °C
Density (g/mL)	2.28 at 25 °C	2.8 (commercial)	3.0
Solubility in water (µg/L)	13.3 (commercial)	1 at 25 °C (commercial)	<0.1
Log Kow	6.57 (commercial)	6.29 (commercial)	6.265
Log Koc	4.89-5.10	5.92-6.22	6.80
Vapor pressure (mmHg)	3.5×10^{-7} (commercial)	4.9×10^{-8} (commercial)	3.2×10^{-8}

Table 6. Some physical and chemical properties of BDE-100

Common Name CAS n.º Molecular formula Molecular weight (g/moL)	Chemical Structure	Solubility in water (µg/L) [a]	Vapor pressure (mmHg) [a]	Henry's law constant (atm. m^3/mol) [a]	Log K_{ow} [b]
BDE-100 189084-64-8 $C_{12}H_5Br_5O$ 564.7	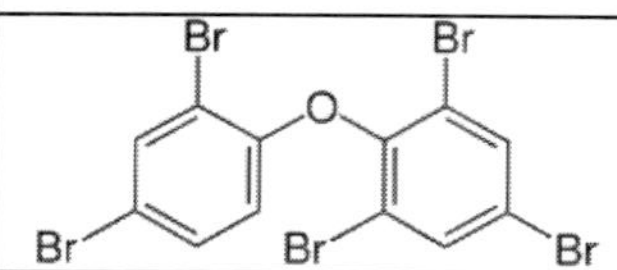	40 (at 25 °C)	2.15×10^{-7}	6.80977×10^{-7}	7.24

Notes:. [a] – [70]; [b] – [71].

Table 7. Studies found in the literature concerning the photodegradation studies of BDE-100 in liquid mediums

Degradation type	PBDE congeners	Initial Concentration	Matrix	Operating conditions	Taken volumes	Extraction/ clean-up	Analytical method	Results and comments	Ref.
Photodegradation	**BDE-100**	5 μg/L	H_2O (ice and liquid)	**Freezer process:** in a laboratory freezer at -20 °C. **Ice solid samples dimensions:** 6 cm diameter and 0.8 cm height. **Radiation source:** two 8 W low-pressure Hg lamps (254 nm). **Irradiation:** maximum 10 min	22 mL	**HS-SPME** Sample volume – 10 mL Fiber – PDMS Extraction temperature – 100 °C Extraction time – 45 min	GC-MS	**Kinetics:** 1st order reaction **Rate constants:** 10.68 h^{-1} (ice), 9.72 h^{-1} (water) **Main mechanism:** consecutive reductive debromination and intramolecular elimination of HBr. **Degradation products:** lower brominated PBDEs and PBDFs.	[73]
Photodegradation	**BDEs 28, 47, 99, 100, 153 and 183**	10 μg/L	Hexane	**Radiation source:** Hg 500 W filtered with Pyrex glass **Irradiation time:** 40 h	The necessary for injection	**Not applicable**	GC-ECD	**Kinetics:** 1st order reaction **Rate constants:** ranging from 0.10 h^{-1} (BDE-100) to 2.64 h^{-1} (BDE-183) **Main mechanism**: consecutive reductive debromination **Photoreactivity of bromines**: decrease from *ortho* to *para* positions for less brominated PBDEs; no differences were observed for higher brominated congeners. **Degradation products:** lower brominated PBDEs.	[66]

Table 7. (Continued)

Degradation type	PBDE congeners	Initial Concentration	Matrix	Operating conditions	Taken volumes	Extraction/ clean-up	Analytical method	Results and comments	Ref.
Redutive debromination by zerovalent iron	**BDEs 209, 100, 66, 47, 28 and 7**	50 mg/L	ethyl acetate	1 mL of a 50 mg/L PBDE solution in ethyl acetate was added to 5 g of iron. Solvent was removed and 10 mL of water was added. Shaken for 3 h to 40 days at 30 °C.	----------	Both aqueous and soil phase were washed three times (20 mL × 3) with a mixture of ether and dichloromethane (1:1). After combined the organic extracts were concentrated to 1 mL.	GC-ECD and HRGC-HRMS	**Kinetics:** 1st order reaction **Rate constants:** 92% of BDE-209 was transformed into lower bromo congeners after 40 days. **Main mechanism**: Debromination is stepwise with bromines being sequencially substituted by hydrogen. **Photoreactivity of bromines**: *Meta* and *ortho*-bromines were more susceptible to debromination than those at the *para* position.	[72]

Notes: ECD – electron capture detector; GC – gas chromatography; HRGC-HRMS – high resolution gas chromatography–mass spectrometry; HS-SPME – headspace solid phase microextraction; MS – mass spectrometry detector.

Toxicity information about PBDEs is scarce and the toxicity assessment of each congener in particular has not been finished yet [55, 63]. Some studies have pointed out their potential to act as hormone disruptors, neurodevelopmental toxic and carcinogenic agents [55, 56]. However, it is inappropriate to make generalized statements about PBDEs toxicology because their physicochemical properties vary considerably from congener to congener [52]. As referred in the beginning of this chapter, we've mainly focused in BDE-100 congener to represent the "new" chemicals. The main properties of this specific congener are presented in Table 6, where the extremely low water solubility should be stressed.

1.3.1. Degradation Technologies for BDE-100 in Waters

The degradation of PBDEs in liquids is a relatively recent research area and, for that reason, only few studies are available in the literature. Most of them are based on experiments carried out in solvents different from pure water to overcome some troubles associated to their extremely low water solubility. However, concerns about the adequacy of these studies to predict the degradation of such chemical compounds in water are emerging. Indeed, the degradation mechanisms and the by-products formed in the reaction are really dependent on the conditions employed.

As indicated previously, BDE-100 was selected to represent the "new" chemical compounds with disastrous effects even at low concentration levels. For that reason, a detailed literature overview was made for this compound in particular. Up to the author's knowledge, only three studies were found about BDE-100 degradation by AOPs and they were not focused only in this congener but also in other ones (Table 7). The first study was published in 2005 and was focused in the degradation of six polybrominated diphenyl ethers (BDEs 7, 28, 47, 66, 100 and 209) with zerovalent iron [72]. It was shown that the degradation occurs by a sequential debromination mechanism being *meta* and *ortho*-bromines more susceptible than those at the *para* position. The degradation of BDE-100 follows a first order kinetic model, being attained only 65% of conversion after 40 days of reaction ($[BDE\text{-}100]_0$ = 50 mg/L in ethyl acetate, Iron = 5 g/mL, T = 30 °C) [72]. Later, Fang et al. (2008) studied the degradation of other six PBDE congeners (BDEs 28, 47, 99, 100, 153 and 183) in hexane under ultraviolet irradiation [66].

It was reported that 55% of the initial BDE-100 was degraded after 8 h of irradiation ($[BDE\text{-}100]_0$ = 10 μg/L in hexane, 500 W mercury lamp) [66]. They also demonstrated that BDE-100 exhibits lower photodegradation rates than the expected ones due to its bromine substitution pattern [66]. In 2012, Sanchez-Prado et al. investigated the degradation of BDE-100 in ice solid samples and in aqueous solutions (containing small amounts of isooctane) using an artificial UV light source [73]. For the experiments carried out in aqueous solutions, 79% of BDE-100 degradation was recorded after exposition to UV light during 10 min ($[BDE\text{-}100]_0$ = 5 μg/L in ethyl acetate, two 8 W low-pressure mercury lamps – 254 nm) [73]. Concerning the reaction by-products, the degradation of BDE-100 generally conducts to the formation of lower brominated PBDE congeners by consecutive reductive debromination [66, 72]. However, other relevant degradation pathways may occur and are responsible for the formation of other classes of compounds, such as the polybrominated dibenzofurans (PBDFs) [73]. As demonstrated, besides the scarce information concerning BDE-100 degradation in liquid phase, the results from the available studies are very distinct. The differences could be attributed to the nature of the removal technology applied and to the different reaction conditions implemented, including UV light sources, initial concentrations of BDE-100 and photolytic matrices employed in the photodegradation studies. In this chapter some preliminary experiments about the degradation of BDE-100 in pure water will be shown (cf. section 4). The great advantage of this study, over the others available in the literature, is the fact that is applicable to waters without the presence of any organic solvents, which is particularly important namely in the context addressed herein. The other reports were based on the degradation of BDE-100 in organic solvents [66] or in aqueous solutions prepared from organic solvent solutions [72, 73].

2. Advanced Oxidation Technologies for the Treatment of Waters Contaminated with Paraquat

In this section, different approaches for the treatment of waters contaminated with PQ will be presented, whose development took into account their applicability in a deliberate contamination scenario of DWDSs. Off-line and on-line approaches will be addressed and correspond to the treatment of the water out and inside of the pipelines, respectively. For off-line treatment, three different processes have been proposed: a homogeneous degradation process by Fenton´s (i) or photo-Fenton's reagent (ii) and a heterogeneous Fenton´s oxidation by using deposits from DWDS as catalysts in the process (iii). Concerning the on-line (or in-situ) treatment, the degradation of PQ in waters was also evaluated by the Fenton's process.

2.1. Off-line Approaches

2.1.1. Paraquat Degradation in Water by Homogeneous Fenton and Photo-Fenton Processes

Herein, it will be presented the most relevant conclusions of a study about PQ degradation in water by oxidation with Fenton´s reagent, done by our research group [42]. Additionally, some results about PQ degradation in water by the photo-Fenton process will be also presented for comparison purposes.

To assess the effect of some variables on the PQ degradation performance in water by the Fenton's reagent, a parametric study was conducted changing one variable at a time. The variables studied were: temperature, iron salt (catalyst) concentration, hydrogen peroxide (oxidant) concentration, initial pH, iron salt source, initial PQ concentration and the mode of oxidant addition. Runs were carried out in a batch reactor [42].

The effect of the temperature on the PQ degradation and on the total organic carbon (TOC) removal, i.e., mineralization degree reached, was evaluated performing experiments at discrete temperatures in the range of 10 to 70 °C (10, 30, 50 and 70 °C) and under the following conditions: $[H_2O_2] = 3.4\times10^{-2}$ M, $[Fe^{2+}] = 5.0\times10^{-4}$ M, $pH_0 = 3.0$ and $[PQ] = 3.9\times10^{-4}$ M (Figure 2). It was concluded that the temperature has a positive effect on the PQ degradation and on the TOC removal, as expected and predicted by the Arrhenius' law (rate constants increase exponentially with temperature). It was also evidenced that an increase of the temperature from 50 to 70 °C had a minor impact on the degradation rate of PQ, which was attributed to the thermal decomposition of the hydrogen peroxide at high temperatures [74, 75]. For the highest temperature tested (70 °C), nearly all the initial PQ present in the reactor was degraded after 60 min and 60% of the organic matter was totally mineralized into CO_2 and H_2O after 240 min (Figure 2).

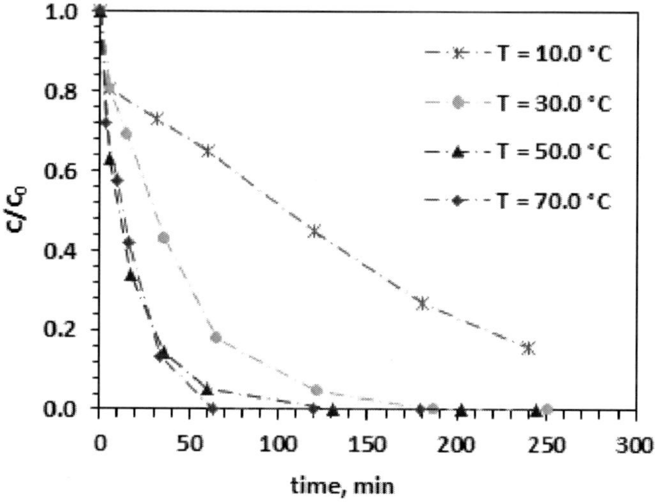

(a)

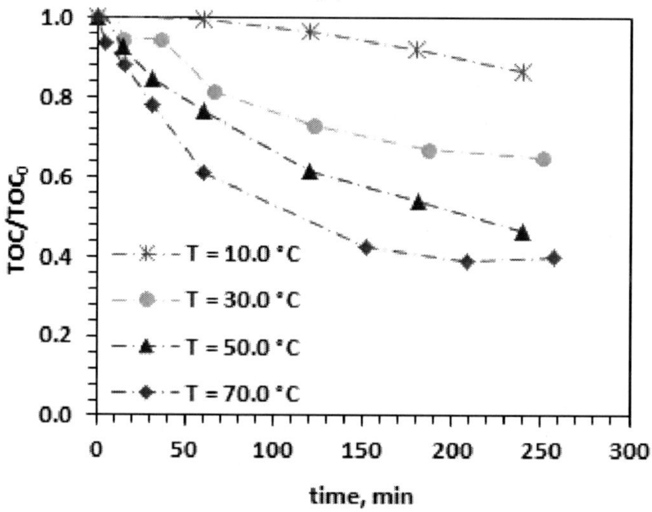

(b)

Figure 2. Temperature effect on the PQ concentration evolution (a) and on the TOC removal (b) as a function of time ($[H_2O_2]_0 = 3.4\times10^{-2}$ M, $[Fe^{2+}]_0 = 5.0\times10^{-4}$ M, $pH_0 =$ 3.0 and $[PQ]_0 = 3.9\times10^{-4}$ M). Dashed lines are merely illustrative of the data trend [42]. Reprinted from Chemical Engineering Journal, volume 175, 279-290, Mónica S. F. Santos, A. Alves, Luís M. Madeira, Paraquat removal from water by oxidation with Fenton's reagent. Copyright (2011), with permission from Elsevier.

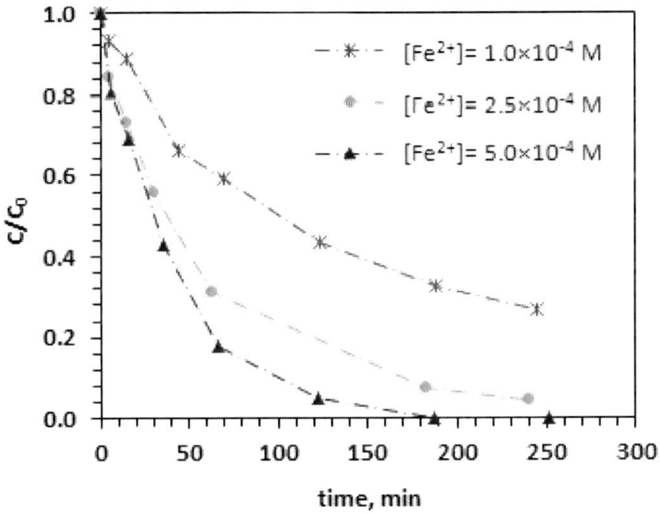

(a)

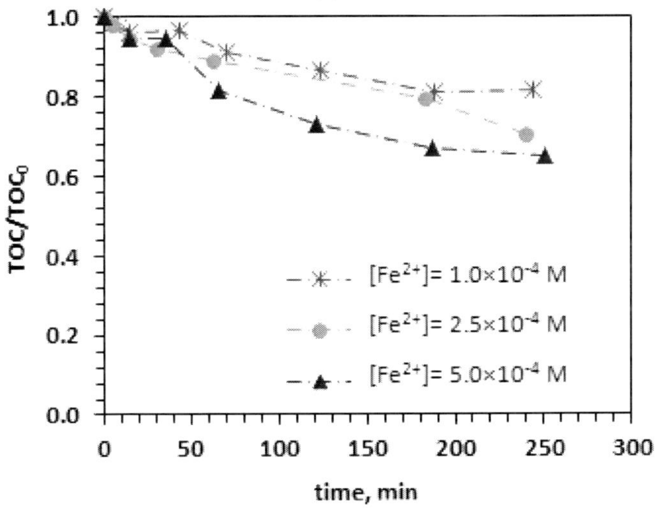

(b)

Figure 3. Fe^{2+} concentration effect on the PQ degradation (a) and on the TOC removal (b) as a function of time (T = 30.0 °C, pH_0 = 3.0, $[H_2O_2]_0 = 3.4\times10^{-2}$ M, and $[PQ]_0 = 3.9\times10^{-4}$ M). Dashed lines are merely illustrative of the data trend [42]. Reprinted from Chemical Engineering Journal, volume 175, 279-290, Mónica S. F. Santos, A. Alves, Luís M. Madeira, Paraquat removal from water by oxidation with Fenton's reagent. Copyright (2011), with permission from Elsevier.

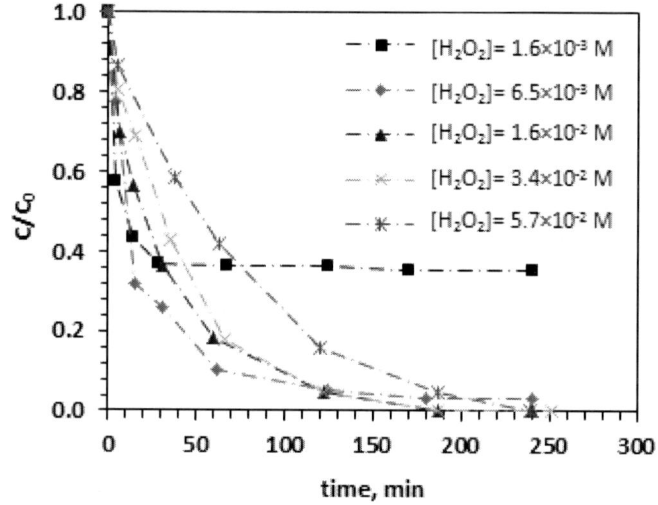

(a)

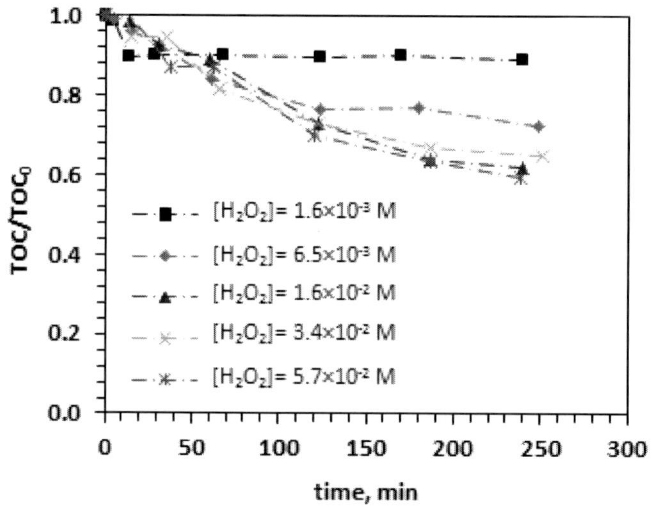

(b)

Figure 4. H_2O_2 concentration effect on the PQ degradation (a) and on the TOC removal (b) as a function of time (T = 30.0 °C, pH_0 = 3.0, $[Fe^{2+}]_0 = 5.0\times10^{-4}$ M, and $[PQ]_0 = 3.9\times10^{-4}$ M). Dashed lines are merely illustrative of the data trend [42]. Reprinted from Chemical Engineering Journal, volume 175, 279-290, Mónica S. F. Santos, A. Alves, Luís M. Madeira, Paraquat removal from water by oxidation with Fenton's reagent. Copyright (2011), with permission from Elsevier.

Attending on the results achieved (and on the energy consumption in real practice), it was proposed to work at 30 °C because only a slight improvement on the PQ degradation profile was attained for higher temperatures.

Concerning the iron salt concentration, the effects of three iron (II) concentrations ($[Fe^{2+}] = 1.0\times10^{-4}$ M, 2.5×10^{-4} M and 5.0×10^{-4} M) on the PQ degradation and on the TOC removal were checked at the following conditions: T = 30 °C, $pH_0 = 3.0$, $[H_2O_2] = 3.4\times10^{-2}$ M and $[PQ] = 3.9\times10^{-4}$ M (Figure 3). For the lowest catalyst dose employed (1.0×10^{-4} M), both the parent compound degradation and mineralization proceed at low rate. As expected, a positive effect on both the PQ degradation and mineralization rates was noticed raising the Fe^{2+} concentration above this value because the salt act as catalyst in the process. For the highest iron concentration studied (5.0×10^{-4} M), complete PQ decomposition (within the analytical uncertainties) is reached after 180 min and 35% is converted into CO_2 and H_2O after the same reaction time (Figure 3).

The effect of the hydrogen peroxide was evaluated for five concentrations in the range of 1.6×10^{-3} M – 5.7×10^{-2} M at T = 30 °C, $pH_0 = 3.0$, $[Fe^{2+}] = 5.0\times10^{-4}$ M and $[PQ] = 3.9\times10^{-4}$ M (Figure 4). The results showed that there is an optimum H_2O_2 concentration (ca. 6.5×10^{-3} –1.6×10^{-2} M).

For the lowest H_2O_2 concentration (1.6×10^{-3} M), the herbicide concentration remains constant after 30 min of reaction because all hydrogen peroxide is consumed, although a rapid initial PQ degradation was observed. This is also observed after 180 min of reaction for an initial dose of 6.5×10^{-3} M (Figure 4). Increasing further the initial H_2O_2 concentration, the initial PQ degradation rate decreases, and for a H_2O_2 dose of 5.7×10^{-2} M, the initial PQ degradation rate is the smallest one. On the other hand, the highest mineralization occurs for this hydrogen peroxide dose (5.7×10^{-2} M). Such behavior resides in the scavenging effect that becomes more significant at high hydrogen peroxide doses and/or to the fact that the oxidant species are preferentially consumed in the oxidation of PQ by-products rather than in the parent compound. For $[H_2O_2] = 1.6\times10^{-2}$ M, 95% of PQ is degraded after 120 min, but for the same time only 27% of PQ is totally mineralized into CO_2 and H_2O (a value of 38% was reached after 240 min) (Figure 4). Additionally, it was proved that the more divided is the hydrogen peroxide addition the better because the scavenging effects are not as significant as when all H_2O_2 is added at once [42].

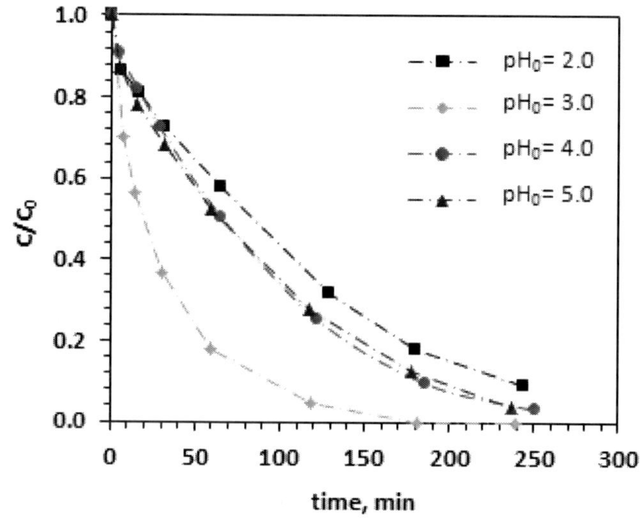

(a)

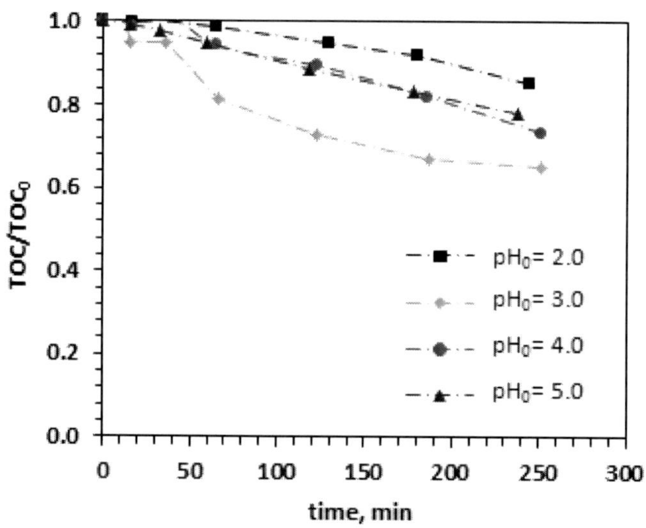

(b)

Figure 5. Initial pH effect on the PQ degradation (a) and on the TOC removal (b) as a function of time (T = 30.0 °C, $[Fe^{2+}]_0 = 5.0\times10^{-4}$ M, $[H_2O_2]_0 = 1.6\times10^{-2}$ M, and $[PQ]_0 = 3.9\times10^{-4}$ M). Dashed lines are merely illustrative of the data trend [42]. Reprinted from Chemical Engineering Journal, volume 175, 279-290, Mónica S. F. Santos, A. Alves, Luís M. Madeira, Paraquat removal from water by oxidation with Fenton's reagent. Copyright (2011), with permission from Elsevier.

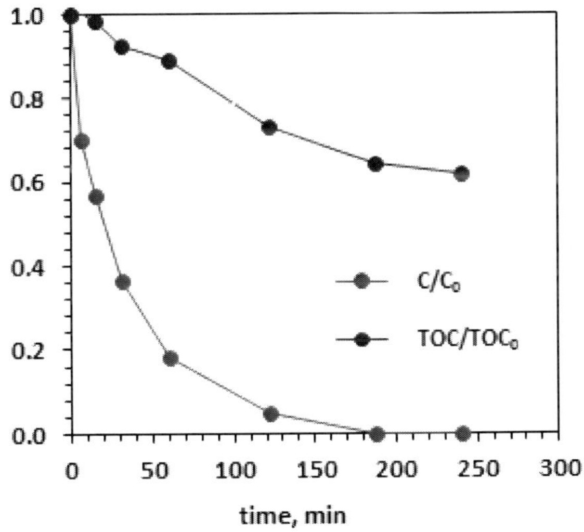

Figure 6. Degradation of paraquat in water by Fenton´s reagent (T = 30 °C, $[Fe^{2+}]$ = 5.0×10^{-4} M, $[H_2O_2] = 1.6\times10^{-2}$ M, $pH_0 = 3.0$ and [PQ] = 3.9×10^{-4} M).

The effect of the initial pH was assessed for pH_0 = 2.0, 3.0, 4.0 and 5.0 (at T = 30 °C, $[H_2O_2] = 1.6\times10^{-2}$ M, $[Fe^{2+}] = 5.0\times10^{-4}$ M and [PQ] = 3.9×10^{-4} M - Figure 5). The results showed that the highest PQ degradation was attained for pH_0 = 3.0, which was explained by the fact that when pH is higher than 3.0, ferrous and ferric oxyhydroxides are formed which inhibit the reaction between Fe^{2+} and H_2O_2 and thus the hydroxyl radicals production. In addition, hydrogen peroxide stability is also smaller at high pH values. On the other hand, at lower pHs excessive H^+ reacts with H_2O_2 to produce $H_3O_2^+$, which is stable and cannot react with iron (II) to form the $HO^\bullet$ species and the hydroxyl radicals can also be scavenged (Figure 5). Even so, for reactions conducted at initial pHs higher than 3.0 the degradation of PQ still remains effective, although at a much slower rate (Figure 5), because the pH of the effluent decreases along the reaction due to the formation of acidic degradation by-products reaching values around 3.0. This is an interesting result in the context of a deliberate or accidental contamination event of drinking water distribution systems because it proves that PQ could be degraded in water without any initial pH adjustment at off-line or even at on-line (in-situ) conditions.

Other parameters and their effects on the PQ degradation performance and on the mineralization degree were also examined (e.g., the iron salt source and the contaminant charge of the effluent) [42], but the results will not be indicated herein for brevity reasons. In sum, this detailed parametric study allowed to understand the PQ degradation profile in waters by oxidation with Fenton's reagent under different reaction conditions and allowed to propose the more adequate conditions for being implemented, for instance upon a deliberate or accidental contamination event of drinking water distribution systems. Attending on the results obtained in terms of degradation and mineralization performances and on practical aspects (for example easiness of operation and energy consumption), the following conditions were proposed: T = 30 °C, $[Fe^{2+}] = 5.0\times10^{-4}$ M, $[H_2O_2] = 1.6\times10^{-2}$ M, $pH_0 = 3.0$ and [PQ] = 3.9×10^{-4} M. Under these conditions, all PQ (analytical standard) was degraded after 4 h of reaction and 40% of the organic matter was totally mineralized into CO_2 and H_2O (Figure 6).

The degradation of the PQ commercial product (Gramoxone - GMX) was also evaluated under those conditions and was compared with that obtained for PQ analytical standard [42]. It was evidenced that the Gramoxone degradation rate was lower (27% mineralization after 323 min versus 38% after 240 min) due to the presence of other organic compounds that also consume the hydrogen peroxide/hydroxyl radicals. Even so, the mineralization degree achieved for PQ analytical standard at the above mentioned conditions was relatively low and, for that reason, the generated effluents were carefully analyzed by LC-DAD and LC-MS techniques to identify the degradation by-products. Three degradation products, resulting from the treatment of PQ-contaminated waters by Fenton's reagent, were detected and identified after 4 h of reaction; they correspond to oxalic acid, isonicotinic acid and 4-carboxy-1-methylpyridinium ion [76]. Additionally, five other molecular ions were detected by LC-MS (*m/z* 201, 265, 267, 283 and 291), but none of them has been matched to chemical structures [76]. It was estimated that the identified products correspond to 61% of the overall remaining organic matter after 4 h of reaction. Furthermore, oxalic acid and isonicotinic acid are less toxic than the PQ molecule [76].

To verify how much the use of UV-Vis light improves the PQ degradation in water, an experiment in the presence of radiation was performed under the same conditions tested for the dark Fenton´s reagent: T = 30 °C, $[Fe^{2+}] = 5.0\times10^{-4}$ M, $[H_2O_2] = 1.6\times10^{-2}$ M, $pH_0 = 3.0$ and [PQ] = 3.9×10^{-4} M. A very fast PQ degradation in water by the photo-Fenton process was achieved, being 99% of PQ degraded after only 5 min of reaction (Figure 7a). The TOC removal was also very high when compared with the Fenton´s reagent (Figure 7b), allowing reaching 96% of mineralization after only one hour of reaction. These interesting results are extremely attractive evidencing the fact that photo-Fenton can be a good alternative to the Fenton's reagent for the off-line treatment of waters contaminated with PQ.

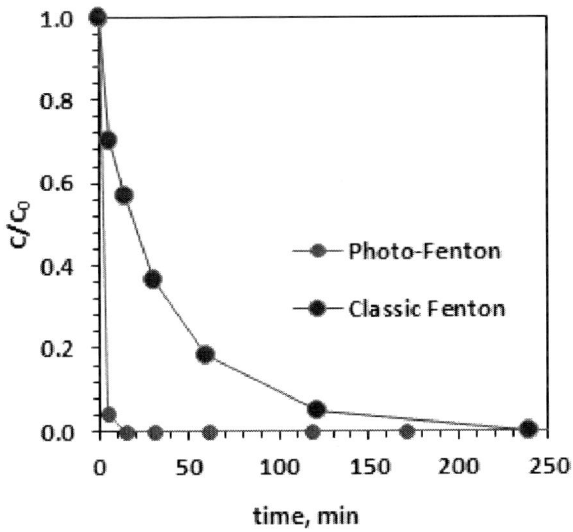

(a)

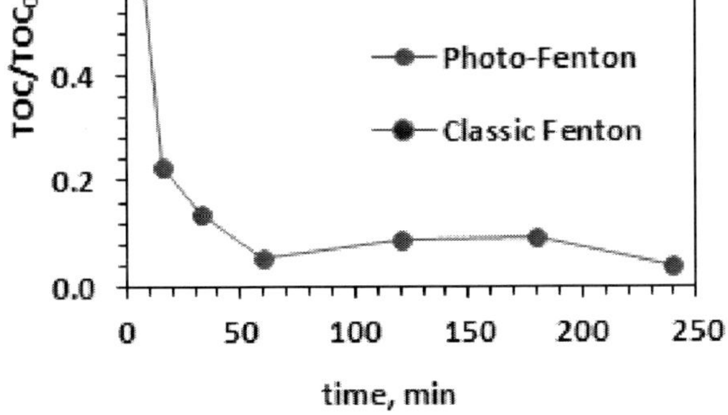

(b)

Figure 7. PQ degradation (a) and mineralization (b) profiles in water by Fenton and photo-Fenton (Heraeus TQ-150 W; 5.9×10^{-7} Einstein/s) processes.

2.1.2. Paraquat Degradation in Water by the Heterogeneous Fenton Process

In order to reduce the amounts of chemicals used in the process (namely iron salts), and trying to avoid the formation of iron sludge, the possibility of using iron-containing materials, or solids with iron attached to its surface, has been widely addressed by several authors through the so-called heterogeneous Fenton-like (or photo-Fenton-like) processes [77-79]. In most cases these materials are used to support the iron species, being worth of mentioning the use of clays [79-83], activated carbons [84-86], or zeolites [79, 87-90]. Besides, catalyst separation and recovery is also promoted with this approach, which can be also extended to naturally occurring iron-containing materials.

Deposits that exist in water networks are mainly of two types: loosely or firmly attached. They are thus often discarded, either after networks cleaning (loosely deposits), or during maintenance operations (when firmly attached). Iron-containing deposits found in water networks were tested in PQ degradation by Fenton's oxidation, as an alternative to the addition of iron salt [91]. Therefore, the solids valorization may be envisaged using this approach. This way the formation of iron sludge is avoided and the pipe deposits, abundant and with no economic value, can be used as catalysts for off-line applications in waste water treatment. In addition, in the event of a network contamination, the deposits present in the pipes can be directly used to catalyze the process in-situ, without requiring the insertion of further chemicals (apart from the oxidant – H_2O_2) , as it will be shown in section 2.2.

Four real deposits, collected from old cast iron pipes that needed to be substituted, with different chemical compositions and morphological properties were characterized and tested as catalysts in PQ heterogeneous peroxidation, aiming also trying to correlate solid's properties/composition with their catalytic performance [91]. For the deposit with better performance a study of the pH effect was made, because this is a variable of paramount importance in the designed applications of DWDS remediation; particular emphasis was given for exploring the possibility of operating in non-acidic conditions that are commonly optimal for Fenton's oxidation.

The elemental composition of each deposit sample tested, herein called S1 to S4, was obtained by the inductively coupled plasma-optical emission spectrometry technique (ICP-OES); results obtained are presented in Table 8. All the deposits are quite complex, with numerous elements in their composition. As it can be seen from Table 8, the deposits S1, S2 and S3 have a high content in aluminum, calcium, copper, silicon and particularly in iron; on the other hand, the deposit S4 is highly rich in calcium.

Table 8. Composition of the deposit samples used (determined by ICP-OES) and corresponding BET areas (determined by N_2 adsorption) and pH_{pzc} values [91]

Element	Samples Composition (mg/kg of deposit, dry basis)			
	S1	S2	S3	S4
Al	1610	1240	882	413
Ca	3860	3010	996	352000
Cd	46.0	1.8	13.5	0.2
Co	27.7	32	27.0	1.9
Cr	15.7	13.2	75.2	1.0
Cu	1130	986	145	72.8
Fe	471000	408000	637000	2910
K	<790	<400	<600	<300
Mg	<790	322	<600	2990
Mn	126	123	4390	894
Na	<790	<400	<600	1060
Ni	42.3	39.9	183	6.3
Pb	33.4	8.8	40.3	2.4
P	315	345	7830	102
Si	2600	1090	1970	819
Zn	92.2	74	2630	47.6
Water content, wt. %	29.2	35.4	30.0	11.4
S_{BET}, m^2/g	7	5	36	1
pH_{pzc}	2.85	3.33	5.99	8.60

The deposits were also analyzed by x-ray diffraction. The search was mainly focused on iron-based oxides, because iron is the most active metal for Fenton-like processes and in the majority of the solids it is the predominant one; however, there are many different allotropic forms of iron oxides-hydroxides, which can be transformed easily between them.

The broad and low intensity peaks pointed out the low crystallinity of the residues S1, S2 and S3 [91]. These deposits are basically a mixture of iron oxyhydroxides, in the case of S1 corresponding to the ferroxyhyte, δ-FeO(OH). Additionally, akaganéite, the Fe^{3+} oxide-hidroxide/chloride ($Fe^{3+}O(OH,Cl)$) that can be formed by the chlorine used in drinking water networks, was identified. The presence of quartz was also found. The XRD peaks of deposit S2 corresponded to lepidocrocite, γ-FeO(OH), and goethite, α-FeO(OH), or $Fe_2O_3 \cdot H_2O$. In the deposit S3 goethite was easily identified. The most pure and crystalline sample was the deposit S4 and, for that reason, the easiest to identify the existing components. The peaks clearly matched calcite, i.e., calcium carbonate – $CaCO_3$ [91].

According to Echeverría et al. [92], deposits from DWDS can be classified in three different categories, attending on their composition: brown, tubercle and white deposits. Brown deposits are composed by aluminosilicates, humic acids and quartz. However, tubercles are mainly comprised by magnetite and goethite and in some cases lepidocrocite. White deposits are essentially composed by calcite, aluminosilicates and quartz. From this information, the identification of each deposit used was done [91]: deposits S1 and S2 were assigned to be included in the group of brown deposits, whereas S3 was identified as a tubercle deposit, and S4 clearly a white deposit. So, deposits S1, S2, S3 and S4 are typical and representative of those found in drinking water distribution systems.

In such a work it was compared the PQ degradation and mineralization performance for each deposit (taking into account the iron content in each one, given in Table 8, i.e., adjusting the mass of catalyst so that the total iron load was the same – 1.0×10^{-3} M) [91]. Figure 8a indicates that deposits S3 and S4 did not lead to a good pesticide degradation (actually with sample S3 no decline in PQ concentration was noticed) in opposition to deposits S1 and S2. Additionally, deposit S1 presented the best degradation degree, being possible to completely eliminate PQ after ca. 2.5 h of reaction. With this sample, a TOC reduction above 40% was obtained after 6 h (Figure 8b). To better clarify the reasons why deposits S3 and S4 did not work as catalysts, some issues were highlighted. First, one should remark that according to the XRD analyses reported above, the presence of calcite in deposit S4 was detected, and inherently of carbonate/bicarbonate ions in solution, which are known for their scavenging activity towards hydroxyl radicals, as it can be seen in Eqs. 8 – 9 [93]. This explained the very low activity of such material [91].

$$HCO_3^- + OH^\cdot \rightarrow CO_3^- + H_2O \qquad (8)$$

$$CO_3^{2-} + OH^\cdot \rightarrow CO_3^- + HO^- \qquad (9)$$

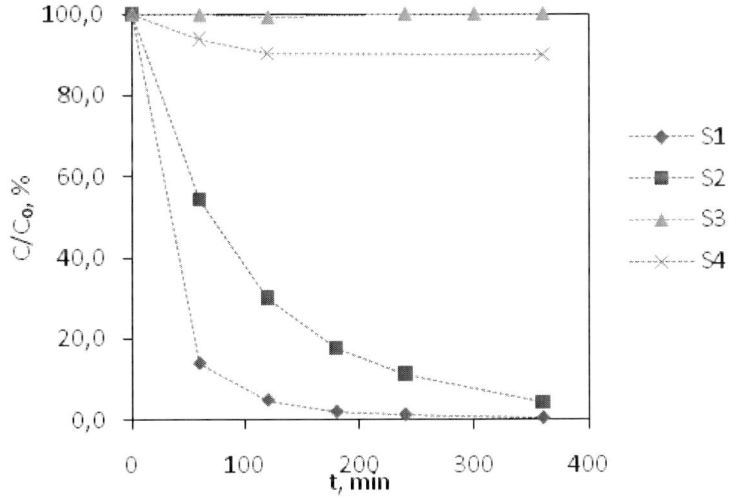

(a)

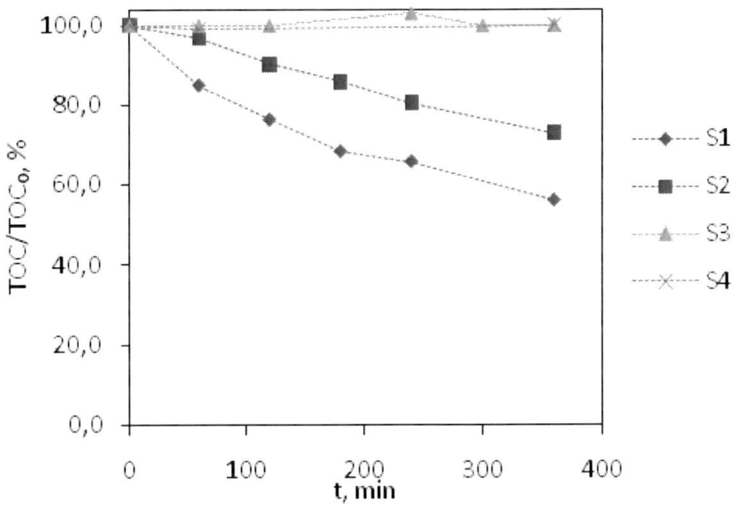

(b)

Figure 8. Paraquat degradation (a) and mineralization (b) along the reaction using the different deposits ([S1] = 118.6 mg/L, [S2] = 136.8 mg/L, [S3] = 87.6 mg/L, [S4] = 19192 mg/L) without pH re-adjustment after solid addition (T = 30 °C, $pH_0 = 3$, $[PQ]_0 = 3.98 \times 10^{-4}$ M, $[H_2O_2]_0 = 1.5 \times 10^{-2}$ M and $[Fe] = 1.0 \times 10^{-3}$ M); dashed lines are merely illustrative of the data trend [91]. Reprinted from Chemical Engineering Journal, volume 210, 339-349, Cátia Oliveira, Mónica S. F. Santos, F. J. Maldonado-Hódar, Gabriela Schaule, Arminda Alves, Luís M. Madeira, Use of pipe deposits from water networks as novel catalysts in paraquat peroxidation. Copyright (2012), with permission from Elsevier.

Deposit S3 has a very high iron content (Table 8), which is not corroborated by its catalytic performance. In fact, as above mentioned, it had no activity at all. This solid is composed by goethite, and according to Matta et al. [94], goethite has a very low catalytic activity in Fenton-like processes compared to other minerals, what can be responsible for the inactivity of this deposit. We should highlight that an experiment carried out with goethite revealed in fact a very low activity in this process and no PQ mineralization at all. On the other hand, sample S3 is the solid with the highest content of Mn (Table 8), a metal that is well-known to decompose hydrogen peroxide, not yielding the radicals [95, 96].

In sum, it was concluded that iron-rich deposits found in DWDS, particularly the ones pertaining to the brown deposits category, can be valorized through their reuse as catalysts in water treatment process such as Fenton´s reaction. Such alternative allows avoiding the use of other chemicals apart from the oxidant, which is particularly relevant when high quantities of water need to be treated, and reduces significantly the formation of iron sludges. Moreover, it opens the door for in-situ applications, as described in the next section.

2.2. On-line Approaches

2.2.1. Paraquat Degradation in Water by the Fenton's Process

In the previous section, it was shown that pipe deposits found in DWDS can act as catalysts in the Fenton's process. This way, such deposits can be reused (upon cleaning/maintenance operations in water networks), or used as catalysts for in situ water treatment, in case of a contamination event.

In a subsequent stage, the main objective was (i) to demonstrate whether with Fenton´s reaction it is possible to degrade PQ commercial product (GMX – gramoxone) directly in a water distribution system and (ii) to assess the influence of length of cast iron pipe sections and amount of loose deposits on the catalytic process [97]. The study was carried out in a pilot loop to simulate a real water supply system. The pilot loop used in these experiments (Figure 9) was made of polyvinyl chloride (PVC) pipes; however, some sections were replaced (see Table 9), if necessary, by iron pipes that should (hopefully) catalyze the process. The total length of the pilot loop was 28 m and with an internal diameter of 75 mm. Iron pipes were obtained from inner heat supply system. However these pipes were made accordingly to the same standard as drinking water system pipes.

Table 9. Operating conditions for each experiment in the loop system [97]

	RUN #1	RUN #2	RUN #3	RUN #4	RUN #5	RUN #6	RUN #7	RUN #8
$[H_2O_2]_0$, M	1.5×10^{-2}	1.5×10^{-2}	1.5×10^{-2}	1.5×10^{-2}	1.5×10^{-2}	1.5×10^{-2}	**1.5×10^{-2}** (*)	**1.5×10^{-2}** (**)
$[FeSO_4\cdot7H_2O]_0$, M	5.0×10^{-4}	5.0×10^{-4}	5.0×10^{-4}	---	---	---	5.0×10^{-4}	5.0×10^{-4}
$[Fe]_{salt}$, mg/L	139	139	139	---	---	---	139	139
[deposit], mg/L	---	---	---	---	**128**	**744**	---	---
$[Fe]_{deposit}$, mg/L	---	---	---	---	**94**	**606**	---	---
$[PQ]_0$, M	3.9×10^{-4}	3.9×10^{-4}	3.9×10^{-4}	3.9×10^{-4}	3.9×10^{-4}	3.9×10^{-4}	3.9×10^{-4}	3.9×10^{-4}
pH_0	**3.0**	**5.0**	3.0	3.0	3.0	3.0	3.0	3.0
$T_{average}$, °C	≈ 20	≈ 20	≈ 20	≈ 20	≈ 20	≈ 20	≈ 20	≈ 20
Pipes types	PVC	PVC	**PVC + iron, 0.5 m**	**PVC + iron, 13.6 m**	PVC	PVC	PVC	PVC
Iron source	Iron (II) salt	Iron (II) salt	Iron (II) salt + iron pipe	Iron pipe	**Loose deposit A**	**Loose deposit B**	Iron (II) salt	Iron (II) salt

(*) Gradual addition of H_2O_2 (32 mL × 5).

(**) Gradual addition of H_2O_2 [(15 mL × 2) + (30 mL × 2) + 70 mL].

Reprinted from Drinking Water Engineering and Science, volume 7, (in press) C. Oliveira, K. Gruskevica, T. Juhna, K. Tihomirova, A. Alves, L. M. Madeira, Removal of paraquat pesticide with Fenton reaction in a pilot scale water system. Copyright (2014).

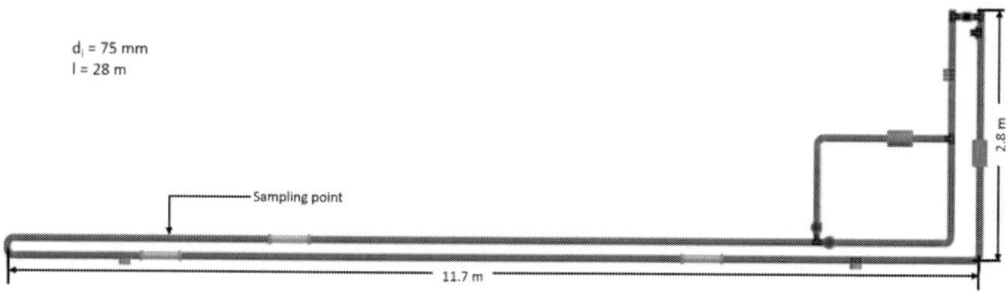

Figure 9. Illustration of the pilot loop used [97]. Reprinted from Drinking Water Engineering and Science, volume 7, (in press) C. Oliveira, K. Gruskevica, T. Juhna, K. Tihomirova, A. Alves, L. M. Madeira, Removal of paraquat pesticide with Fenton reaction in a pilot scale water system. Copyright (2014).

Experiments were performed under the same conditions used in lab scale runs (section 2.1.1), but at room temperature. Indeed, temperature is not controllable neither in the pilot loop nor in a real situation. The operating conditions used for each experiment in the pilot loop (recirculation tubular reactor) are presented in Table 9, where parameters changed in each run are highlighted in bold. The first step was to find out if the degradation process in the pilot loop would be similar to the one in the lab scale, so that it could be easily scaled up to the size of a real water network. This possibility would represent a main novelty and step forward as a decontamination in-situ.

The same operational conditions were then applied in both lab and pilot scale, and compared in terms of PQ degradation and Gramoxone mineralization. Figure 10a presents the changes of PQ concentration in both reactors, i.e., the evolution of the pesticide concentration along reaction time, while Figure 10b refers to the dissolved organic carbon (DOC) data.

It can be seen that the performance of the process, for both degradation and mineralization, was very similar in both reactors, although their dimensions and mode of operation were considerably different, as well as initial DOC (because GMX instead of the PQ analytical standard was used in the pilot loop reactor, as described above) [97]. Besides, it was shown that the decrease of temperature (20 °C in the loop instead of 30 °C) did not influence significantly the process, representing an important advancement in water decontamination in-situ, avoiding the concerns about high temperatures. It should be noted that after 8 h, in the pilot loop, PQ was completely degraded, even using lower temperatures than in the lab scale; it could thus be concluded that the process can be easily scaled up, keeping the good performance.

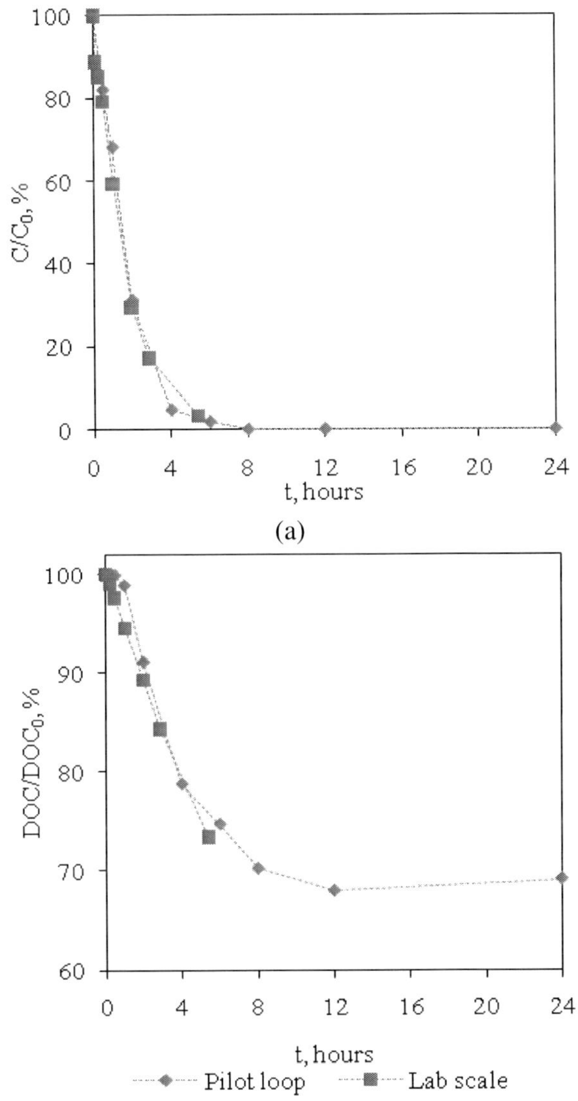

(b)

Figure 10. Comparison of PQ degradation (a) and PQ/GMX mineralization (b) between experiments in lab scale (T = 30 °C) and in the pilot loop (T = 20 °C, for the other conditions refer to Table 9, RUN #1); dashed lines are merely illustrative of the data trend [97]. Reprinted from Drinking Water Engineering and Science, volume 7, (in press) C. Oliveira, K. Gruskevica, T. Juhna, K. Tihomirova, A. Alves, L. M. Madeira, Removal of paraquat pesticide with Fenton reaction in a pilot scale water system. Copyright (2014).

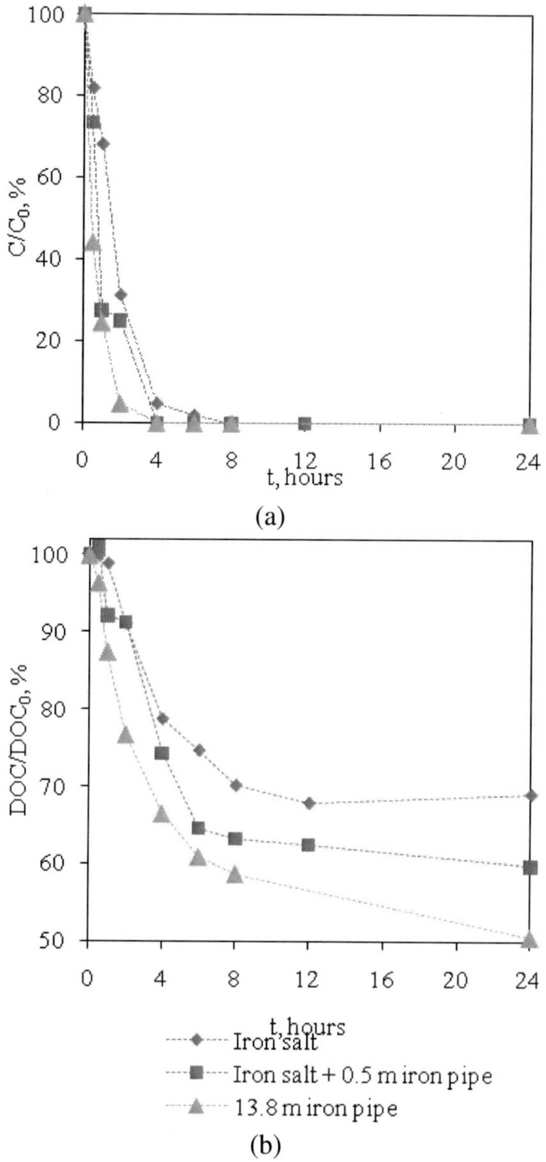

Figure 11. Effect of the use of iron pipes in the PQ degradation (a) and GMX mineralization (b) along the time of reaction (for the conditions refer to Table 9, Runs #1, #3 and #4); dashed lines are merely illustrative of the data trend [97]. Reprinted from Drinking Water Engineering and Science, volume 7, (in press) C. Oliveira, K. Gruskevica, T. Juhna, K. Tihomirova, A. Alves, L. M. Madeira, Removal of paraquat pesticide with Fenton reaction in a pilot scale water system. Copyright (2014).

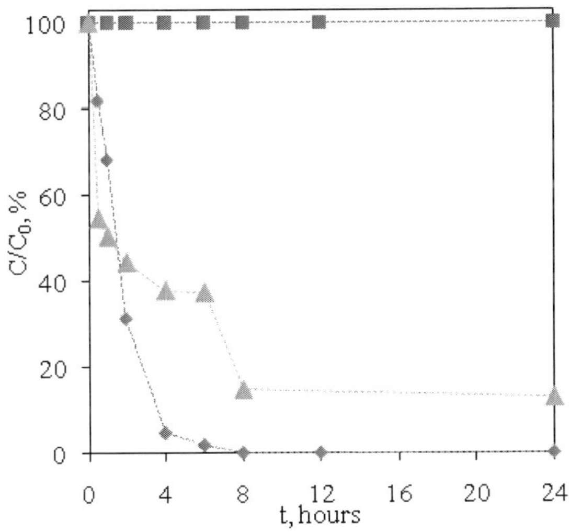

(a)

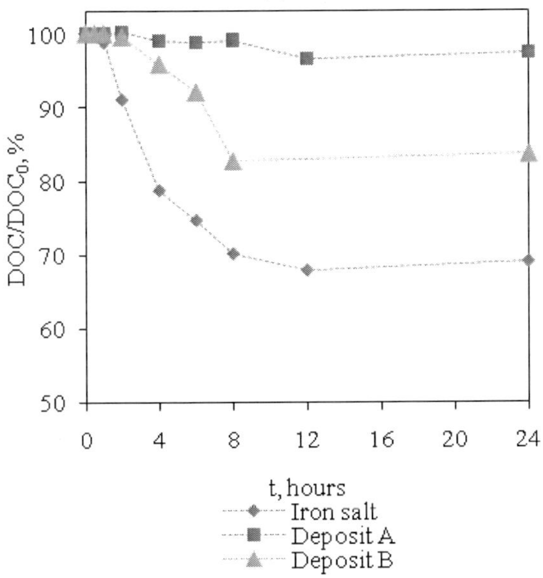

(b)

Figure 12. Effect of the loose deposits in the PQ degradation (a) and GMX mineralization (b) along the time of reaction (Runs #1, #5 and #6); dashed lines are merely illustrative of the data trend [97]. Reprinted from Drinking Water Engineering and Science, volume 7, (in press) C. Oliveira, K. Gruskevica, T. Juhna, K. Tihomirova, A. Alves, L. M. Madeira, Removal of paraquat pesticide with Fenton reaction in a pilot scale water system. Copyright (2014).

To evaluate the ability of pipes from water distribution systems to catalyze the Fenton's reaction, some sections of the PVC from the pilot loop were replaced by used iron pipes (see Table 9) [97]. The installed iron pipe sections represented approximately 2 and 50% of the total length of the loop (RUN #3 and RUN #4, respectively). Results presented in Figure 11a show that the PQ degradation rate, in the experiment where no iron pipes were installed, was slower than the experiment where 2% iron pipes were installed (0.5 m long iron pipe), although also in the presence of $FeSO_4$. This shows that the presence of the iron pipe can promote a faster production of hydroxyl radicals and thus increase the rate of PQ degradation. As it could be predicted, the greater the part of the iron pipes, the faster was the pesticide degradation; this can be confirmed by the analysis of the data from Figure 11a, from the experiment where 50% of PVC pipes were replaced by iron pipes (and no iron salt was used), which shows that the degradation rate is the fastest among those tested. Once the pipes can promote the catalysis, this can be considered as a great advantage in case of a decontamination demand since it allows operating in-situ and avoiding the use of some chemicals. It should be noted that the fastest and highest mineralization was also achieved in the experiment where 50% of the loop consisted of iron pipes (Figure 11b). According to the data shown in Figure 11a and Figure 11b, these conditions provided a more effective degradation, with complete pesticide degradation and a remarkable mineralization of 50% (such mineralization was due not only to the carbon present in the pesticide PQ, but also to that in other organics present, as the pesticide employed was not analytical grade but rather commercial gramoxone) [97]. Thus, it can be stated that the heterogeneous reaction is quite effective for PQ degradation and mineralization of organics.

The thickness of the iron pipe used in RUN #3 (2% of pipes made of iron) was checked before and after the experiment [97]. This measurement was done using ultrasound. The results showed that pipe wall thickness decreased 10542 nm during the 48 h of reaction, meaning that pipe wall thickness decrease rate was approximately 1.92 mm/year. In normal conditions, for the iron pipe, the thickness reduction along time would be around 160 nm/day. Using more aggressive conditions against pipes permanently, i.e., like the ones employed for in situ decontamination (pH = 3 and $[H_2O_2] = 1.5 \times 10^{-2}$ M), one can conclude that this pipe would need to be replaced only after 3-4 years of use [97].

Two loose deposit samples (one obtained from a real drinking water distribution system – deposit A – and another obtained from the tower of a water distribution system – deposit B) were tested for their ability to catalyze the Fenton's reaction in the same pilot loop [97]. Both deposits were mainly a mixture of iron oxyhydroxides. RUN #1 shows the degradation process in the presence of iron salt, while RUN #5 and RUN #6 present, respectively, the oxidation in the presence of the deposit A and the deposit B. The iron content was 73.4% in deposit A and 81.5% in deposit B, which means that the iron concentration used was, respectively, 94 mg Fe/L and 606 mg Fe/L – cf. Table 9. In RUN #1, an iron concentration of 5.0×10^{-4} M (139 mg Fe/L) was provided by the addition of $FeSO_4$. Analyzing Figure 12a and Figure 12b it can be said that the process was more effective when using $FeSO_4$; the experiment with deposit A yielded no PQ degradation after 24 h, while deposit B showed oxidation performance, being able to almost completely degrade PQ. Significant difference in performances between deposits cannot be explained by the different doses used, but rather by their natures and composition. This highlights again the possibility of using some pipe deposits as catalysts of the Fenton's process, which performance depends on the deposit used, in agreement with the results presented in section 2.1.2.

The absence of catalytic activity of deposit A can be due to the fact of being mostly composed by $CaCO_3$ and goethite, which, according to Matta et al. [94] and Oliveira et al. [91], has a very low catalytic activity when compared with other iron minerals.

3. Advanced Oxidation Technologies for the Treatment of Waters and Solids Contaminated with Chlorfenvinphos

3.1. Chlorfenvinphos Degradation in Water by the Fenton Process

In order to study the effect of the main operating conditions in the CFVP degradation in waters, a detailed parametric study was performed [98] and the main results are herein described. The parametric study performed consisted in the assessment of the influence of several concentrations of H_2O_2 (oxidizing agent) or Fe^{2+} (catalyst) in the degradation degree, as well as the effect of the initial pH and medium temperature.

It was found that the variation in the concentration of H_2O_2 or Fe^{2+} affect positively or negatively the degree of CFVP degradation and mineralization, once high dosages can participate in parallel reactions, acting as scavengers in the process; low doses slow down the process [98]. For instance, when increasing the oxidant dose from 0.008 to 0.015 M, the rate of CFVP degradation was improved because this led to an increase in the generation of radical species responsible for the degradation of the pesticide. Nevertheless, when H_2O_2 concentrations were in the range of 0.015 M to 0.022 M, the degree of degradation achieved was very similar, suggesting the possibility of occurring parallel reactions between the excess of H_2O_2 and hydroxyl radicals (scavenging effect), as described by eq. 3 [98].

Different Fe^{2+} concentrations were also tested [98]. In the experiments using lower iron concentrations it was observed a slower degradation, due to the insufficient amount of catalyst for the same amount of oxidant (Figure 13). It is evident that the best and faster degradation is the one using an initial concentration of Fe^{2+} of 4.6×10^{-4} M.

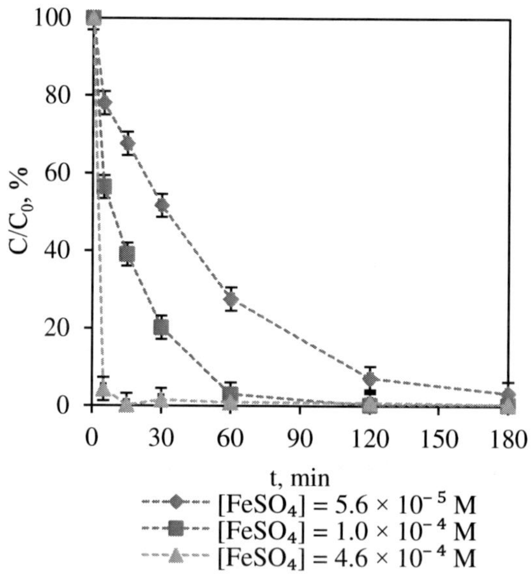

Figure 13. Effect of the iron concentration in the CFVP degradation in water ($[CFVP]_0 = 2.8 \times 10^{-4}$ M, $[H_2O_2]_0 = 1.5 \times 10^{-2}$ M, $pH_0 = 3.0$, T = 30.0 °C); dashed lines are merely illustrative of the data trend [98]. Reprinted from Chemical Engineering Journal, volume 241, 190-199, Cátia Oliveira, Arminda Alves, Luís M. Madeira, Treatment of water networks (waters and deposits) contaminated with chlorfenvinphos by oxidation with Fenton's reagent. Copyright (2014), with permission from Elsevier.

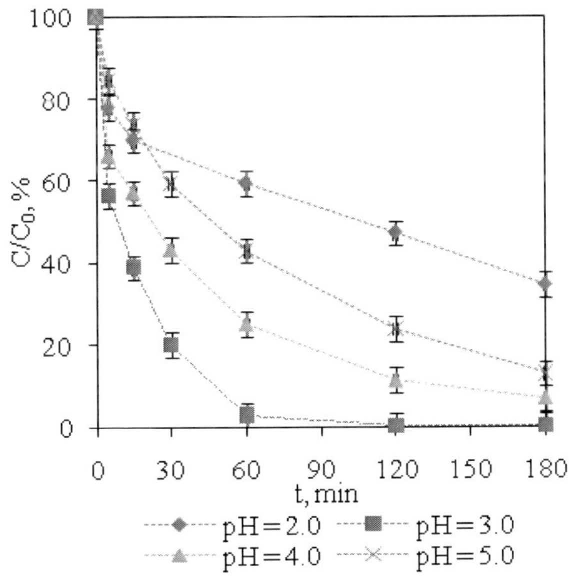

Figure 14. Effect of the initial pH in degradation CFVP in water ($[CFVP]_0 = 2.8 \times 10^{-4}$ M, $[H_2O_2]_0 = 1.5 \times 10^{-2}$ M, $[Fe^{2+}]_0 = 4.6 \times 10^{-4}$ M, T = 30.0 °C); dashed lines are merely illustrative of the data trend [98]. Reprinted from Chemical Engineering Journal, volume 241, 190-199, Cátia Oliveira, Arminda Alves, Luís M. Madeira, Treatment of water networks (waters and deposits) contaminated with chlorfenvinphos by oxidation with Fenton's reagent. Copyright (2014), with permission from Elsevier.

One of the most important factors in the Fenton's reaction is the initial pH used in the process, as shown in Figure 14. Several experiments were performed using a different pH, in the range from 2 to 5 [98]. Figure 14 shows that the degradation was most effective when an initial pH of 3.0 was used, once after 1 h of reaction about 97% degradation of the parent compound was achieved. On the other hand, under these conditions, at the beginning of the reaction there was a high decrease of the CFVP concentration of about 40% in only 5 minutes. The worst performance at lower or higher pH values is ascribed to the same reasons described above for PQ – cf. section 2.1.1.

Temperature can also affect the CFVP degradation/mineralization degree; some experiments were performed to test the influence of this variable [98]. It was found that the faster degradation occurs for higher temperatures, as commonly found out in the literature. This trend could be explained by the influence of temperature in the kinetic constants according to Arrhenius law [99]. For a regular temperature (~30 °C), a relatively fast degradation could be achieved, although the degradation can be much faster for higher temperatures.

Additionally, the mineralization of CFVP became more efficient as the temperature of the medium increased [98].

To assess the behavior of some solids as catalysts in the degradation of CFVP in water by the Fenton´s process, three deposits were tested [98]. The experiments and comparisons were performed considering the iron content of each solid (pipe deposit, kaolin and goethite). Kaolin was found to be an inert, once any CFVP degradation was found; this suggested that the iron present in kaolin is not able to react with the hydrogen peroxide for producing the hydroxyl radicals. Goethite showed a low catalytic activity and the pipe deposit S2 (already studied in section 2.1.2) presented more than 90% of degradation after a short period of time (4 h) [98].

In short, it can be concluded that if the DWDS are contaminated with any compound belonging to a chemical group represented by PQ and CFVP, the conditions employed for the treatment of the water by Fenton´s reagent under off-line should be similar ($[Fe^{2+}]_0 = 5.0\times10^{-4}$ M, $[H_2O_2]_0 = 1.6\times10^{-2}$ M, pH = 3 and T = 30 °C). Additionally, it was also proved that, as demonstrated for PQ, brown deposits can be used as catalysts in the process avoiding the addition of other iron sources and reducing the iron sludge. Moreover, it can be advanced that the on-line treatment, making use of the active iron available in deposits and pipe materials, could also be a reliable alternative for the treatment of waters contaminated with CFVP, as occurs for PQ.

3.2. Chlorfenvinphos Degradation in Solids by the Heterogeneous Fenton Process

After contamination of a DWDS, solids present therein (i.e., in the pipe walls or fine loose deposits) can be also contaminated. So, in a recent work it was envisaged the possibility of degrading CFVP in both solid and liquid contaminated phases, in a slurry batch reactor [98]. To assess the effect of the nature of the iron source, different samples were used as contaminated solid (kaolin, goethite, and kaolin blended with a pipe deposit). For kaolin, one decontamination experiment was done in the presence of a $FeSO_4$ solution 4.6×10^{-4} M, a second was performed in the presence of a $FeSO_4$ solution 1.1×10^{-2} M and a third experiment was conducted only using the iron present in the kaolin as catalyst in the reaction; thus, in total, 5 runs were carried out with different sources of iron, either in solution or from the solids. The pesticide concentration in each phase, liquid or solid, and conversions obtained along the time of decontamination are shown in Figure 15 and in Figure 16.

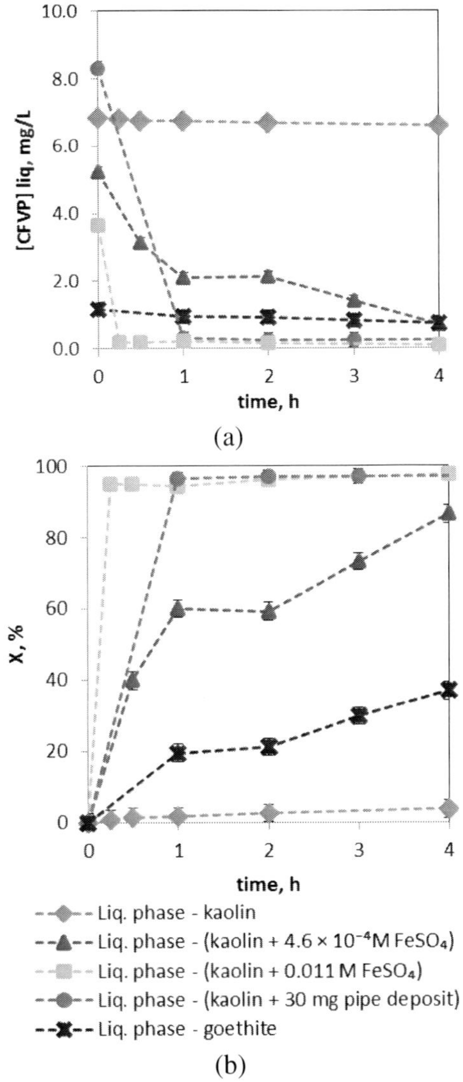

Figure 15. CFVP concentration (a) and conversion (b) in the liquid phase along the reaction for the different contaminated materials submitted to Fenton's reaction ($[H_2O_2]_0 = 1.5 \times 10^{-2}$ M, $pH_0 = 3.0$, $T \approx 20$ °C, $[Fe]_{kaolin} = 2.5 \times 10^{-2}$ M, $[Fe]_{(kaolin + 4.6 \times 10^{-4} M FeSO_4)} = 2.6 \times 10^{-2}$ M and $[Fe]_{(kaolin + 0.011 M FeSO_4)} = 3.6 \times 10^{-2}$ M, $[Fe]_{(kaolin + pipe deposit)} = 3.6 \times 10^{-2}$ M, $[Fe]_{goethite} = 6.1 \times 10^{-1}$ M); Dashed lines are merely illustrative of the data trend [98]. Reprinted from Chemical Engineering Journal, volume 241, 190-199, Cátia Oliveira, Arminda Alves, Luís M. Madeira, Treatment of water networks (waters and deposits) contaminated with chlorfenvinphos by oxidation with Fenton's reagent. Copyright (2014), with permission from Elsevier.

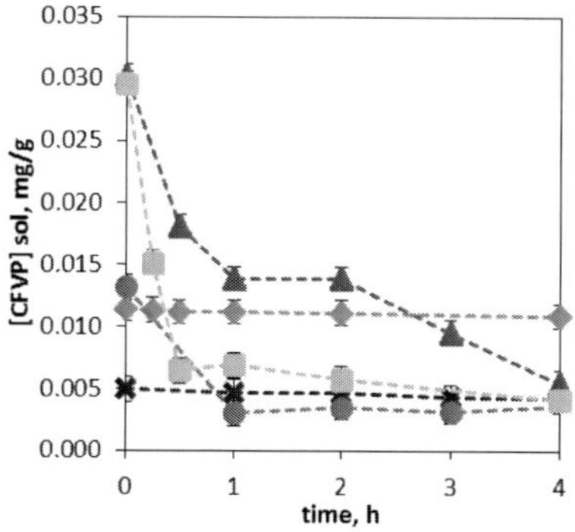

(a)

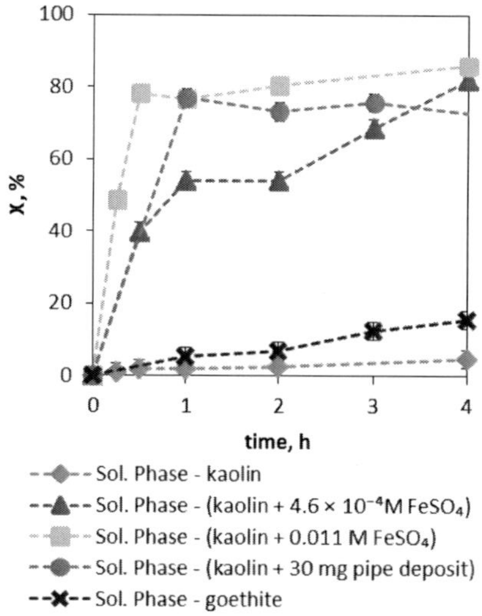

(b)

Figure 16. CFVP concentration (a) and conversion (b) in the solid phase along the reaction for the different contaminated materials submitted to Fenton's reaction ($[H_2O_2]_0 = 1.5 \times 10^{-2}$ M, $pH_0 = 3.0$, T ≈ 20 °C, $[Fe]_{kaolin} = 2.5 \times 10^{-2}$ M, $[Fe]_{(kaolin + 4.6 \times 10^{-4} M FeSO_4)} = 2.6 \times 10^{-2}$ M and $[Fe]_{(kaolin + 0.011 M FeSO_4)} = 3.6 \times 10^{-2}$ M, $[Fe]_{(kaolin + pipe deposit)} = 3.6 \times 10^{-2}$ M, $[Fe]_{goethite} = 6.1 \times 10^{-1}$ M); Dashed lines are merely illustrative of the data trend [98]. Reprinted from Chemical Engineering Journal, volume 241, 190-199, Cátia Oliveira, Arminda Alves, Luís M. Madeira, Treatment of water networks (waters and deposits) contaminated with chlorfenvinphos by oxidation with Fenton's reagent. Copyright (2014), with permission from Elsevier.

According to the results shown in Figure 15a and Figure 16a, kaolin itself did not work as catalyst and consequently CFVP concentrations kept constant along the decontamination step, either in the liquid or in the solid. To decontaminate kaolin, it was necessary to add an extra source of iron (iron sulphate in solution, in this case). When using an iron sulphate concentration of 4.6×10^{-4} M, the pesticide was almost completely degraded in the liquid phase, with an overall conversion > 80% after 4 h (Figure 15); curiously, a similar pattern and performance was noticed in the solid (Figure 16). Using a $FeSO_4$ dose of 1.1×10^{-2} M, the degradation was tremendously accelerated, putting into evidence the role of the iron salt as catalyst (CFVP conversion was above 90% in the liquid after less than 30 min, reaching in the solid also a very good performance – conversion of 80% in the same decontamination period). The mixture of kaolin with the pipe deposit was also tested and it was found that the decontamination was very effective, removing more than 95% of CFVP from the liquid phase and about 75% in the solid phase in only 1 h (Figure 15 and Figure 16) [98]. It should be highlighted that in this experiment it was used a total iron concentration equivalent to the one employing kaolin and a dissolved $FeSO_4$ salt dose of 1.1×10^{-2} M (total iron dosage = 3.6×10^{-2} M). Although with the solids only, the process is slightly slower as compared to the one using the dissolved iron salt, which is related to the availability of iron in heterogeneous vs. homogeneous process, final conversions are quite similar in both cases, reinforcing that the iron from the pipe deposit is available to react with the hydrogen peroxide, being its use effective as catalyst in the decontamination of the solids. This result was remarkable, showing an important advancement in solid decontamination, because some natural iron-containing deposits can be easily decontaminated, using the iron present in their own composition as catalyst. Data in Figure 15 and Figure 16 also show that goethite can be used as catalyst, but the conversion values are very low (around 35% in the liquid and 15% in the solid phase after 4 h).

4. Advanced Oxidation Technologies for the Treatment of Waters Contaminated with BDE-100

4.1. Degradation of BDE-100 in Waters by Photolysis

As mentioned above, studies focused on the use of advanced oxidation processes towards BDE-100 degradation are scarce, and most make use of radiation-assisted processes in presence of organic solvents.

The degradation of BDE-100 in water by direct photolysis was thus studied, at different light intensities. First, solutions of BDE-100 (50 μg/L) were directly irradiated by a 150 W medium-pressure vapor lamp placed inside a quartz jacket, where water circulation was promoted by a thermostatic bath for temperature control. Then, the water flowing through the quartz jacket was substituted by two different solutions with different concentrations of a dye (Vert Cibacrone T3G-E), this way allowing to change the light intensity, i.e., the photonic flux, reaching the BDE-100 solution to degrade. The effect of the photonic flux on the BDE-100 degradation in water is depicted in Figure 17 (dye concentrations were approximately constant during the degradation experiments, and so a constant photonic flux was ensured).

As demonstrated in Figure 17, the intensity of the light is an important factor on the BDE-100 degradation by this methodology. None BDE-100 degradation was attained for a photonic flux of 1.3×10^{-9} Einstein/s. For the maximum light intensity, nearly 80% of BDE-100 was degraded after 80 min of reaction, reaching 92% after 6 h.

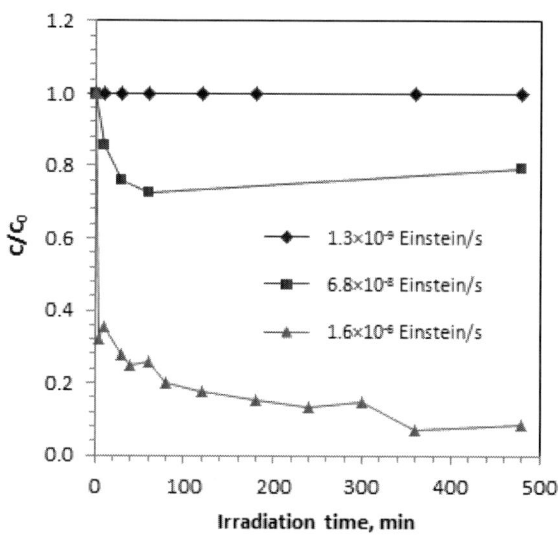

Figure 17. BDE-100 degradation in water by direct photolysis using different light intensities ($[BDE\text{-}100]_0$ = 50 μg/L; pH_0= 3.0 and T=25±3 °C).

So, it can be advanced that the degradation methodology proposed here conducts to attractive results in terms of BDE-100 degradation rates, since a very quick decay was noticed in only 5 min of reaction. As explained before, the great advantage of this study over the others available in the literature is the fact that this is applicable to pure water, i.e., without the presence of organic solvents. This is particularly important because using conditions as much as possible closer to the reality, it is easier to predict the efficacy of the chemical compounds degradation in waters by a certain oxidation process.

CONCLUSION

Two different approaches (off-line and on-line) are discussed for PQ degradation in water, which might be required and implemented after contamination of a drinking water network. The dark classic Fenton, making use of hydrogen peroxide and an iron salt as catalyst, was the first removal technology optimized and implemented to degrade PQ outside the DWDS (off-line approach). Under selected conditions (T = 30 °C, $[Fe^{2+}] = 5.0\times10^{-4}$ M, $[H_2O_2] = 1.6\times10^{-2}$ M, $pH_0 = 3.0$ and $[PQ] = 3.9\times10^{-4}$ M), complete PQ degradation and 40% mineralization were reached after 4 h of reaction. The PQ degradation rate in water and, particularly, the mineralization degree, were significantly enhanced under the same conditions, but in the presence of UV light (96% of mineralization after 1 h of reaction). These results suggest that the use of the photo-Fenton process in off-line mode has a tremendous advantage over the Fenton´s reaction if the mineralization degree represents the main objective and if the process has to be accelerated.

It was also addressed the degradation of PQ in waters by the heterogeneous Fenton process using iron-containing loose deposits from DWDS as catalysts. This approach proved to be an innovative alternative. Furthermore, it provides a useful application of these solid wastes that result from either cleaning or maintenance operations in water distribution systems. PQ degradation in water by homogeneous Fenton´s reagent was also done in a pilot loop system, simulating a real DWDS, achieving similar results to those obtained in a lab scale reactor, i.e., this process is effective in both scales. Additionally, it was shown that the cast iron segments placed into the PVC pipe loop system and the loose deposits can work as catalysts in the Fenton´s reaction process, which can thus be a promising technology for in-situ remediation.

As shown for PQ, some iron-rich pipe deposits can also be used as catalysts for the degradation of CFVP in water by the heterogeneous Fenton´s reagent. Additionally, solids contaminated with CFVP can be also submitted to this advanced oxidation process, once it was found to be possible to decontaminate the solids making use of their active iron as catalyst source. In sum, some iron-rich deposits are thus efficient for this decontamination strategy, allowing in-situ remediation of both the liquid and contaminated solid in water networks.

The degradation of BDE-100 in pure water was tested by direct photolysis. Although very preliminary, promising results were obtained being evident a rapid decline of the BDE-100 concentration in the first 5 min of exposition to a UV light with intensity of 1.6×10^{-6} Einstein/s. For this photonic flux, nearly 80% of BDE-100 was degraded after 80 min of reaction, reaching 92% after 6 h.

In sum, it was shown that advanced oxidation technologies can be used with success, in different conditions, for the emergency treatment of waters and deposits from DWDS upon a deliberate (or accidental) contamination event.

Acknowledgments

This work was undertaken as part of the European Research Project SecurEau (http://www.secureau.eu/ – Contract no. 217976), supported by the European commission within the 7th Framework Programme FP7-SEC-2007-1 - “Security; Increasing the security of Citizen; Water distribution surveillance”.

M. Santos is grateful to CCDRN-ON2 for her postdoctoral grant (FEUP-ON2-25-AA-Ciencias Tecnologias Ambiente). The authors are also grateful to FCT for the financial support through the project PTDC/AAC-AMB/101687/2008.

References

[1] WHO, Hazards in drinking-water supply and waste management, in: Health aspects of plumbing, 2006.

[2] J. Fawell and A. D. Hulsmann, Health effects of chemical contamination of drinking water supplies, in: Water and Health.

[3] P. H. Gleick, G. H. Wolff, H. Cooley, M. Palaniappan, A. Samulon, E. Lee, J. Morrison and D. Katz, The World's Water 2006-2007: The Biennial Report on Freshwater Resources, Island Press, 2006.

[4] P. H. Gleick, Water and terrorism, *Water policy* 8 (2006) 481-503.

[5] SecurEau, http://www.secureau.eu/, retrieved on 05-10-2013.
[6] Y. Picó, G. Font, J. C. Moltó and J. Mañes, Solid-phase extraction of quaternary ammonium herbicides, *J. Chromatogr. A* 885 (2000) 251-271.
[7] Paraquat Information Center, http://paraquat.com/, retrieved on 06-06-2013.
[8] H. J. Beckie, Herbicide-resistant weed management: focus on glyphosate, Pest Manag. Sci. 67 (2011) 1037-1048.
[9] T. W. Eubank, D. H. Poston, V. K. Nandula, C. H. Koger, D. R. Shaw and D. B. Reynolds, Glyphosate-resistant Horseweed (Conyza canadensis) Control Using Glyphosate-, Paraquat-, and Glufosinate-Based Herbicide Programs, *Weed Technol.* 22 (2008) 16-21.
[10] D. de Souza, L. Codognoto, S. A. S. Machado and L. A. Avaca, Electroanalytical determination of the herbicide paraquat in natural water and commercial tea samples with gold electrodes obtained from recordable compact disc, *Anal. Lett.* 38 (2005) 331-341.
[11] I. C. Lopes, D. de Souza, S. A. S. Machado and A. A. Tanaka, Voltammetric detection of paraquat pesticide on a phthalocyanine-based pyrolitic graphite electrode, *Anal. Bioanal. Chem.* 388 (2007) 1907-1914.
[12] N. Mai, X. Liu, W. Wei, S. Luo and W. Liu, Electrochemical determination of paraquat using a DNA-modified carbon ionic liquid electrode, *Microchim. Acta* 174 (2011) 89-95.
[13] G. C. Schmitt, C. Paniz, D. Grotto, J. Valentini, K. L. Schott, V. J. Pomblum and S. C. Garcia, Aspectos gerais e diagnóstico clinicolaboratorial da intoxicação por paraquat, *J. Bras. Patol. Med. Lab.* 42 (2006) 235-243.
[14] L. R. Goldfrank and N. E. Flomenbau, Goldfrank's Toxicological Emergencies, 6th ed., McGraw-Hill Professional Publishing, 1998.
[15] A. M. Tsatsakis, K. Perakis and E. Koumantakis, Experience with acute paraquat poisoning in Crete, *Vet. Hum. Toxicol.* 38 (1996) 113-117.
[16] C. M. Chen and A. C. Lua, Lung toxicity of paraquat in the rat, *J. Toxicol. Environ. Health A.* 60 (2000) 477-487.
[17] G. R. Sagar, Uses and Usefulness of Paraquat, *Hum. Exp. Toxicol.* 6 (1987) 7-11.
[18] B. G. Stephens and S. K. Moormeister, Homicidal Poisoning by Paraquat, *Am. J. Forensic Med. Pathol.* 18 (1997) 33-39.

[19] C. D. Klaassen, M. O. Amdur and J. Doull, Casarett and Doull's Toxicology: The basic science of poisons, Macmillan Publishing Company, New York, 1986.

[20] M. A. El Mhammedi, M. Bakasse and A. Chtaini, Electrochemical studies and square wave voltammetry of paraquat at natural phosphate modified carbon paste electrode, *J. Hazard. Mater.* 145 (2007) 1-7.

[21] R. J. Dinis-Oliveira, F. Remião, H. Carmo, J. A. Duarte, A. S. Navarro, M. L. Bastos and F. Carvalho, Paraquat exposure as an etiological factor of Parkinson's disease, *NeuroToxicology* 27 (2006) 1110-1122.

[22] R. Franco, S. Li, H. Rodriguez-Rocha, M. Burns and M.I. Panayiotidis, Molecular mechanisms of pesticide-induced neurotoxicity: Relevance to Parkinson's disease, *Chem. Biol. Interact.* 188 (2010) 289-300.

[23] M. A. El Mhammedi, M. Bakasse and A. Chtainia, Square-wave voltammetric determination of paraquat at carbon paste electrode modified with hydroxyapatite, *Electroanal.* 19 (2007) 1727-1733.

[24] U. M. F. de Oliveira, J. Lichtig and J. C. Masini, Evaluation of a nafion coated glassy carbon electrode for determination of paraquat by differential pulse voltammetry, *J. Braz. Chem. Soc.* 15 (2004) 735-741.

[25] M. Ibáñez, Y. Picó and J. Mañes, Improving the solid-phase extraction of 'quat' pesticides from water samples: Removal of interferences, *J. Chromatogr. A* 823 (1998) 137-146.

[26] *Official Journal of the European Union, Council Directive* 98/83/CE, 1998.

[27] S. Esplugas, J. Giménez, S. Contreras, E. Pascual and M. Rodríguez, Comparison of different advanced oxidation processes for phenol degradation, *Water Res.* 36 (2002) 1034-1042.

[28] H. J. H. Fenton, LXXIII.-Oxidation of tartaric acid in presence of iron, *J. Chem. Soc.* 65 (1894) 899-910.

[29] J. H. Sun, S. P. Sun, M. H. Fan, H. Q. Guo, L. P. Qiao and R. X. Sun, A kinetic study on the degradation of p-nitroaniline by Fenton oxidation process, *J. Hazard. Mater.* 148 (2007) 172-177.

[30] C. Jiang, S. Pang, F. Ouyang, J. Ma and J. Jiang, A new insight into Fenton and Fenton-like processes for water treatment, *J. Hazard. Mater.* 174 (2010) 813-817.

[31] K. Tennakone and I. R. M. Kottegoda, Photocatalytic mineralization of paraquat dissolved in water by TiO2 supported on polythene and polypropylene films, *J. Photochem. Photobiol. A-Chem.* 93 (1996) 79-81.

[32] E. Moctezuma, E. Leyva, E. Monreal, N. Villegas and D. Infante, Photocatalytic degradation of the herbicide "Paraquat", *Chemosphere* 39 (1999) 511-517.

[33] M. Kang, Preparation of TiO2 photocatalyst film and its catalytic performance for 1,1'-dimethyl-4,4'-bipyidium dichloride decomposition, *Appl. Catal. B-Environ.* 37 (2002) 187-196.

[34] M. H. Florêncio, E. Pires, A. L. Castro, M. R. Nunes, C. Borges and F. M. Costa, Photodegradation of Diquat and Paraquat in aqueous solutions by titanium dioxide: evolution of degradation reactions and characterisation of intermediates, *Chemosphere* 55 (2004) 345-355.

[35] M. J. Cantavenera, I. Catanzaro, V. Loddo, L. Palmisano and G. Sciandrello, Photocatalytic degradation of paraquat and genotoxicity of its intermediate products, *J. Photoch. Photobio. A* 185 (2007) 277-282.

[36] E. Kanchanatip, N. Grisdanurak, R. Thongruang and A. Neramittagapong, Degradation of paraquat under visible light over fullerene modified V-TiO2, *Reac. Kinet. Mech. Cat.* 103 (2011) 227-237.

[37] P. C. Kearney, J. M. Ruth, Q. Zeng and P. Mazzocchi, UV ozonation of paraquat, *J. Agric. Food Chem.* 33 (1985) 953-957.

[38] R. Andreozzi, A. Insola, V. Caprio and M.G. D'Amore, Ozonation of 1, 1' dimethyl, 4,4' bipyridinium dichloride (Paraquat) in aqueous solution, *Environ. Technol.* 14 (1993) 695-700.

[39] A. Dhaouadi and N. Adhoum, Heterogeneous catalytic wet peroxide oxidation of paraquat in the presence of modified activated carbon, *Appl. Catal. B-Environ.* 97 (2010) 227-235.

[40] A. Dhaouadi and N. Adhoum, Degradation of paraquat herbicide by electrochemical advanced oxidation methods, *J. Electroanal. Chem.* 637 (2009) 33-42.

[41] J.-C. Lee, M.-S. Kim, C. Kim, C.-H. Chung, S. Cho, G. Han, K. Yoon and B.-W. Kim, Removal of paraquat in aqueous suspension of TiO2 in an immersed UV photoreactor, *Korean J. Chem. Eng.* 20 (2003) 862-868.

[42] M. S. F. Santos, A. Alves and L. M. Madeira, Paraquat removal from water by oxidation with Fenton's reagent, *Chem. Eng. J.* 175 (2011) 279-290.

[43] Agency for Toxic Substances & Disease Registry (ATSDR), http://www.atsdr.cdc.gov/toxprofiles/tp83-c4.pdf, retrieved on 14-01-2014.

[44] B. Sosnowska, B. Huras, A. Krokosz and B. Bukowska, The effect of bromfenvinphos, its impurities and chlorfenvinphos on acetylcholinesterase activity, *Int. J. Biol. Macromol.* 57 (2013) 38-44.

[45] S. Sanches, M. T. Barreto Crespo and V. J. Pereira, Drinking water treatment of priority pesticides using low pressure UV photolysis and advanced oxidation processes, *Water Res.* 44 (2010) 1809-1818.

[46] Water Framework Directive 2000/60/EC of the European Parliament and of the Council of 23 October 2000, European Union.

[47] N. Klamerth, W. Gernjak, S. Malato, A. Agüera and B. Lendl, Photo-Fenton decomposition of chlorfenvinphos: Determination of reaction pathway, *Water Res.* 43 (2009) 441-449.

[48] J. L. Acero, F. J. Real, F. Javier Benitez and A. González, Oxidation of chlorfenvinphos in ultrapure and natural waters by ozonation and photochemical processes, *Water Res.* 42 (2008) 3198-3206.

[49] M. Hincapié, M. I. Maldonado, I. Oller, W. Gernjak, J. A. Sánchez-Pérez, M.M. Ballesteros and S. Malato, Solar photocatalytic degradation and detoxification of EU priority substances, *Catal. Today* 101 (2005) 203-210.

[50] M. Lapertot, S. Ebrahimi, S. Dazio, A. Rubinelli and C. Pulgarin, Photo-Fenton and biological integrated process for degradation of a mixture of pesticides, *J. Photochem. Photobiol. A-Chem.* 186 (2007) 34-40.

[51] M. J. Farré, M. I. Franch, S. Malato, J. A. Ayllón, J. Peral and X. Doménech, Degradation of some biorecalcitrant pesticides by homogeneous and heterogeneous photocatalytic ozonation, *Chemosphere* 58 (2005) 1127-1133.

[52] K. D'Silva, A. Fernandes and M. Rose, Brominated Organic Micropollutants—Igniting the Flame Retardant Issue, Crit. Rev. Environ. Sci. Technol. 34 (2004) 141-207.

[53] O. I. Olukunle, O. J. Okonkwo, K. K. Kefeni and M. Lupankwa, Determination of brominated flame retardants in Jukskei River catchment area in Gauteng, South Africa, *Water Sci. Technol.* 65 (2012) 743-749.

[54] Q. Xiao, B. Hu, J. Duan, M. He and W. Zu, Analysis of PBDEs in Soil, Dust, Spiked Lake Water, and Human Serum Samples by Hollow Fiber-Liquid Phase Microextraction Combined with GC-ICP-MS, *J. Am. Soc. Mass Spectr.* 18 (2007) 1740-1748.

[55] I. Fulara and M. Czaplicka, Methods for determination of polybrominated diphenyl ethers in environmental samples - Review, *J. Sep. Sci.* 35 (2012) 2075-2087.

[56] M. Polo, G. Gómez-Noya, J. B. Quintana, M. Llompart, C. García-Jares and R. Cela, Development of a Solid-Phase Microextraction Gas Chromatography/Tandem Mass Spectrometry Method for Polybrominated Diphenyl Ethers and Polybrominated Biphenyls in Water Samples, *Anal. Chem.* 76 (2004) 1054-1062.

[57] E. Eljarrat and D. Barceló, Sample handling and analysis of brominated flame retardants in soil and sludge samples, *Trends Anal. Chem.* 23 (2004) 727-736.

[58] C. A. de Wit, An overview of brominated flame retardants in the environment, *Chemosphere* 46 (2002) 583-624.

[59] P. O. Darnerud, G. S. Eriksen, T. Jóhannesson, P. B. Larsen and M. Viluksela, Polybrominated diphenyl ethers: Occurence, dietary exposure, and toxicology, *Environ. Health Perspect.* 109 (2001) 49-68.

[60] *Official Journal of the European Union, Directive* 2003/11/EC of the European Parliament and of the Council, 2003.

[61] A. Möller, Z. Xie, R. Sturm and R. Ebinghaus, Polybrominated diphenyl ethers (PBDEs) and alternative brominated flame retardants in air and seawater of the European Arctic, *Environ. Pollut.* 159 (2011) 1577-1583.

[62] Stockholm Convention, http://chm.pops.int/Convention/ThePOPs/TheNewPOPs/tabid/2511/Default.aspx, retrieved on 05-10-2013.

[63] A. Bacaloni, L. Callipo, E. Corradini, P. Giansanti, R. Gubbiotti, R. Samperi and A. Laganà, Liquid chromatography–negative ion atmospheric pressure photoionization tandem mass spectrometry for the determination of brominated flame retardants in environmental water and industrial effluents, *J. Chromatogr. A* 1216 (2009) 6400-6409.

[64] M. L. Hardy, The toxicology of commercial PBDE flame retardants: DecaBDE, octaBDE, pentaBDE, *Organohalogen. Comp.* 47 (2000) 41-44.

[65] R. Millischer, F. Girault, R. Heywood, G. Clarke, D. Hossack and M. Clair, Decabromobiphenyl: toxicological study, *Toxicol. Eur. Res.* 2 (1979) 155-161.

[66] L. Fang, J. Huang, G. Yu and L. Wang, Photochemical degradation of six polybrominated diphenyl ether congeners under ultraviolet irradiation in hexane, *Chemosphere* 71 (2008) 258-267.

[67] C. A. de Wit, M. Alaee and D. G. Muir, Levels and trends of brominated flame retardants in the Artic, *Chemosphere* 64 (2006) 209-233.

[68] K. Hooper and T. A. McDonald, The PBDEs: An emerging environmental challenge and another reason for breast-milk monitoring programs, *Environ. Health Perspect.* 108 (2000) 387-392.

[69] ATSDR, Health Effects, in: Toxicological profile for polybrominated biphenyls and polybrominated diphenyl ethers, 2004, pp. 57-343.

[70] S. A. Tittlemier, T. Halldorson, G. A. Stern and G. T. Tomy, Vapor pressures, aqueous solubilities, and Henry's law constants of some brominated flame retardants, *Environ. Toxicol. Chem.* 21 (2002) 1804-1810.

[71] E. Braekevelt, S. A. Tittlemier and G. T. Tomy, Direct measurement of octanol–water partition coefficients of some environmentally relevant brominated diphenyl ether congeners, *Chemosphere* 51 (2003) 563-567.

[72] Y.-S. Keum and Q. X. Li, Reductive Debromination of Polybrominated Diphenyl Ethers by Zerovalent Iron, *Environ. Sci. Technol.* 39 (2005) 2280-2286.

[73] L. Sanchez-Prado, K. Kalafata, S. Risticevic, J. Pawliszyn, M. Lores, M. Llompart, N. Kalogerakis and E. Psillakis, Ice photolysis of 2,2′,4,4′,6-pentabromodiphenyl ether (BDE-100): Laboratory investigations using solid phase microextraction, *Anal. Chim. Acta* 742 (2012) 90-96.

[74] I. Gulkaya, G. Surucu and F. Dilek, Importance of H2O2/Fe2+ ratio in Fenton's treatment of a carpet dyeing wastewater, *J. Hazard. Mater.* 136 (2006) 763-769.

[75] J. H. Ramirez, F. M. Duarte, F. G. Martins, C. A. Costa and L. M. Madeira, Modelling of the synthetic dye Orange II degradation using Fenton's reagent: From batch to continuous reactor operation, *Chem. Eng. J.* 148 (2009) 394-404.

[76] Rúben Lopes Casal Ribeiro, Deteção de subprodutos resultantes da degradação do pesticida paraquato por oxidação química (Detection of paraquat degradation by-products resultants from chemical oxidation), Thesis of Integrated Master in Chemical Engineering, Faculdade de Engenharia da Universidade do Porto, 2012.

[77] F. Duarte, F. J. Maldonado-Hódar, A. F. Pérez-Cadenas and L. M. Madeira, Fenton-like degradation of azo-dye Orange II catalyzed by transition metals on carbon aerogels, *Appl. Catal. B-Environ.* 85 (2009) 139-147.

[78] J. Herney-Ramirez, M. A. Vicente and L. M. Madeira, Heterogeneous photo-Fenton oxidation with pillared clay-based catalysts for wastewater treatment: A review, *Appl. Catal. B-Environ.* 98 (2010) 10-26.

[79] S. Navalon, M. Alvaro and H. Garcia, Heterogeneous Fenton catalysts based on clays, silicas and zeolites, *Appl. Catal. B-Environ.* 99 (2010) 1-26.

[80] E. G. Garrido-Ramírez, B. K. G. Theng and M. L. Mora, Clays and oxide minerals as catalysts and nanocatalysts in Fenton-like reactions — A review, *Appl. Clay Sci.* 47 (2010) 182-192.

[81] J. Feng, X. Hu, P. L. Yue, H. Y. Zhu and G. Q. Lu, Discoloration and mineralization of Reactive Red HE-3B by heterogeneous photo-Fenton reaction, *Water Res.* 37 (2003) 3776-3784.

[82] J. H. Ramirez, C. A. Costa, L. M. Madeira, G. Mata, M. A. Vicente, M. L. Rojas-Cervantes, A. J. López-Peinado and R. M. Martín-Aranda, Fenton-like oxidation of Orange II solutions using heterogeneous catalysts based on saponite clay, *Appl. Catal. B-Environ.* 71 (2007) 44-56.

[83] H. Hassan and B. H. Hameed, Fe–clay as effective heterogeneous Fenton catalyst for the decolorization of Reactive Blue 4, *Chem. Eng. J.* 171 (2011) 912-918.

[84] M. A. Fontecha-Cámara, M. A. Álvarez-Merino, F. Carrasco-Marín, M. V. López-Ramón and C. Moreno-Castilla, Heterogeneous and homogeneous Fenton processes using activated carbon for the removal of the herbicide amitrole from water, *Appl. Catal. B-Environ.* 101 (2011) 425-430.

[85] F. Duarte, F. J. Maldonado-Hódar and L. M. Madeira, Influence of the characteristics of carbon materials on their behaviour as heterogeneous Fenton catalysts for the elimination of the azo dye Orange II from aqueous solutions, *Appl. Catal. B-Environ.* 103 (2011) 109-115.

[86] T. D. Nguyen, N. H. Phan, M. H. Do and K. T. Ngo, Magnetic Fe2MO4 (M:Fe, Mn) activated carbons: Fabrication, characterization and heterogeneous Fenton oxidation of methyl orange, *J. Hazard. Mater.* 185 (2011) 653-661.

[87] F. Duarte and L. M. Madeira, Fenton- and Photo-Fenton-Like Degradation of a Textile Dye by Heterogeneous Processes with Fe/ZSM-5 Zeolite, *Separ. Sci. Technol.* 45 (2010) 1512-1520.

[88] H. Kušić, N. Koprivanac and I. Selanec, Fe-exchanged zeolite as the effective heterogeneous Fenton-type catalyst for the organic pollutant minimization: UV irradiation assistance, *Chemosphere* 65 (2006) 65-73.

[89] M. Tekbaş, H. C. Yatmaz and N. Bektaş, Heterogeneous photo-Fenton oxidation of reactive azo dye solutions using iron exchanged zeolite as a catalyst, *Micropor. Mesopor. Mat.* 115 (2008) 594-602.

[90] M. Dükkancı, G. Gündüz, S. Yılmaz and R. V. Prihod'ko, Heterogeneous Fenton-like degradation of Rhodamine 6G in water using CuFeZSM-5 zeolite catalyst prepared by hydrothermal synthesis, *J. Hazard. Mater.* 181 (2010) 343-350.

[91] C. Oliveira, M. S. F. Santos, F. J. Maldonado-Hódar, G. Schaule, A. Alves and L. M. Madeira, Use of pipe deposits from water networks as novel catalysts in paraquat peroxidation, *Chem. Eng. J.* 210 (2012) 339-349.

[92] F. Echeverría, J. G. Castaño, C. Arroyave, G. Peñuela, A. Ramírez and J. Morató, Characterization of deposits formed in a water distribution system, *Rev. Chil. Ing.* 17 (2009) 275-281.

[93] M.-C. Lu, J.-N. Chen and C.-P. Chang, Effect of inorganic ions on the oxidation of dichlorvos insecticide with Fenton's reagent, *Chemosphere* 35 (1997) 2285-2293.

[94] R. Matta, K. Hanna and S. Chiron, Fenton-like oxidation of 2,4,6-trinitrotoluene using different iron minerals, *Sci. Total Environ.* 385 (2007) 242-251.

[95] M. Wekesa and Y. Ni, Mechanism of hydrogen peroxide decomposition by manganese dioxide, *Tappi J.* 2 (2003) 23-26.

[96] M. Wekesa, A. Habtewold and J. Mirdaniali, Stabilization of manganese (Mn)-induced peroxide decomposition, *Afr. J. Pure Appl. Chem.* 5 (2011) 176-180.

[97] C. Oliveira, K. Gruskevica, T. Juhna, K. Tihomirova, A. Alves and L.M. Madeira, Removal of paraquat pesticide with Fenton reaction in a pilot scale water system, *Drink. Water Eng. Sci. Discuss.* 7 (2014) p. 11-21.

[98] C. Oliveira, A. Alves and L. M. Madeira, Treatment of water networks (waters and deposits) contaminated with chlorfenvinphos by oxidation with Fenton's reagent, *Chem. Eng. J.* 241 (2014) 190-199.

[99] C. S. D. Rodrigues, L. M. Madeira and R. A. R. Boaventura, Optimization of the azo dye Procion Red H-EXL degradation by Fenton's reagent using experimental design, *J. Hazard. Mater.* 164 (2009) 987-994.

In: Drinking Water and Water Management
Editor: Mason Giannino
ISBN: 978-1-63117-523-7

Chapter 3

ANALYTICAL METHODS TO MONITOR HALOACETIC ACIDS IN WATERS INTENDED FOR HUMAN CONSUMPTION

Bárbara Delgado, Priscilla Rocío-Bautista, Verónica Pino* and Juan H. Ayala

Department of Analytical Chemistry, Nutrition and Food Science, University of La Laguna, Spain

ABSTRACT

Water is an essential natural resource for life. Nowadays, water contamination is a common issue due to a variety of sources but mainly of anthropogenic origin, such as urban and industrial wastes.

Chlorination is probably the most common process used worldwide to treat waters intended to human consumption, because it efficiently removes pathogenic microorganisms like those responsible of malaria or typhus. However, the chlorine used in the process can react with the natural organic matter already present in waters, mainly humic and fulvic acids, originating disinfection byproducts (DBPs). The extent of such formation is also dictated by the bromine content of waters as well as by the acidity. Among disinfection byproducts, haloacetic acids (HAAs) and trihalomethanes (THMs) can be cited.

* corresponding author: veropino@ull.edu.es

HAAs are highly polar compounds. Their structure is based in that of acetic acid molecule, but having the alpha carbon adjacent to the carbonyl with one, two or three halogens atoms (Br or Cl). There is a growing concern about the presence of HAAs in waters. Current in vivo studies with mice have proved toxic and carcinogen effect. Indeed, the International Agency of Research on Cancer (IARC) has classified four HAAs as possible human carcinogenic, and the United States Environmental Protection Agency (US-EPA) classifies one HAA (trichloroacetic acid) as possible human carcinogen and another HAA (dichloroacetic acid) as probable human carcinogen.

In this sense, there is a need of analytical methods for the efficient determination of HAAs in water. Current analytical methods take use of gas-chromatography (GC), high-performance liquid chromatography (HPLC), and ion-chromatography (IC), all in combination with a variety of detection systems. When GC is utilized, a derivatization step is needed to increase the volatility of HAAs (forming esters in the majority of cases). Afterwards, the extraction is normally conducted using headspace (HS) or solid-phase microextraction (SPME). When HPLC is used, the most common strategy is to employ hydrophilic interaction liquid chromatography (HILIC), generally in combination with a supported liquid membrane microextraction (SLMME) method. When IC is used (with or without a suppressor column), the extraction strategy normally involves solid-phase extraction (SPE). This book chapter intends to give an overview of the most recent trends in analytical methods to monitor HAAs in waters intended for human consumption, after their chlorination.

1. Formation of Haloacetic Acids During Chlorination

Chlorination of public water supplies is recognized as one of the greatest achievements in the field of public health in the XX century, removing numerous diseases such as cholera, typhus and dysentery.

Towards the middle of the 70s, it was discovered that, in addition to eliminate pathogens, chlorine reacts with organic matter present in water generating disinfection byproducts (DBPs), specifically synthetic organic compounds such as trihalomethanes (THMs)[1]. Later on, other DBPs have been identified, such as haloacetic acids (AHAs) [2, 3]. Figure 1 includes a general scheme regarding their formation. These compounds can be incorporated into drinking waters during the chlorination step but their formation can also occur in the distribution systems of drinking waters [4, 5].

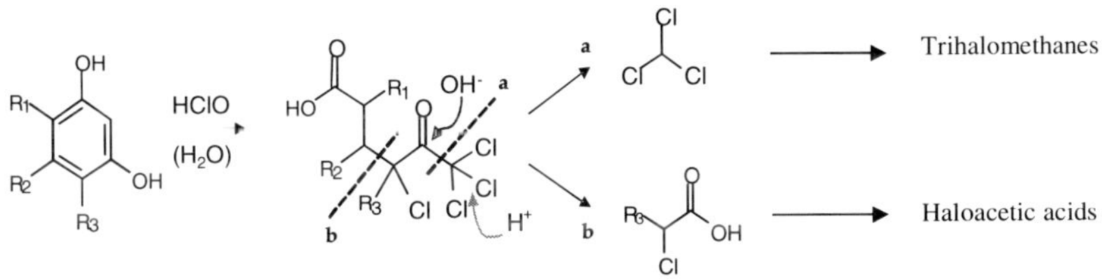

Figure 1. Formation of disinfection byproducts.

Despite the fact that HAAs have not been investigated as thoroughly as THMs, they are probably the main byproducts of water chlorination. The rate and extent of formation of HAAs increase depending on the concentration of chlorine and organic matter, temperature, pH, and concentration of bromine ions in water.

There are 9 haloacetic acids congeners containing chlorine or bromine, as shown in Table 1. For simplicity, haloacetic acids are grouped according to the probability of finding them in chlorinated waters, and the analytical methods used to detect them. Thus, the abbreviation HAA5s includes monochloroacetic acid (MCAA), dichloroacetic acid (DCAA), trichloroacetic acid (TCAA), monobromoacetic acid (MBAA) and dibromoacetic acid (DBAA), and these acids are most commonly studied in waters and subjected to the regulations of the United States Environmental Protection Agency (US-EPA). The abbreviation HAA6s includes all HAAs belonging to HAA5s altogether with bromochloroacetic acid (BCAA). Finally, the abbreviation HAA9s includes all previous HAAs altogether with tribromoacetic acid (TBAA), bromodichloroacetic acid (BDCAA), and chlorodibromoacetic acid (CDBAA). These acids are generally difficult to determine due to their strong acidity: their pKa values oscillated between 0.05 and 2.73, and their hydrophilic character (see Table 1).

2. Health Issues Associated to Haloacetic Acids and Water Policies

Currently, there is a growing concern regarding the adverse effects on human health that some HAAs can cause. The adverse effects of HAAs have been demonstrated with *in vivo* studies [6-9], with evident toxic and carcinogenic effects, as well as damage of the liver, reproduction problems, and alteration of the embryo [10, 11].

Table 1. Main haloacetic acids and their characteristics

Name	Abbreviation	Structure	Abbreviation	Molecular weight ($g \cdot mol^{-1}$)	pK_a
Monochloroacetic acid	MCAA	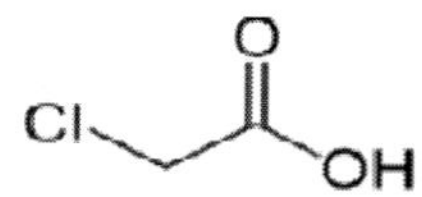	$CH_2ClCOOH$	94.5	2.65
Monobromoacetic acid	MBAA	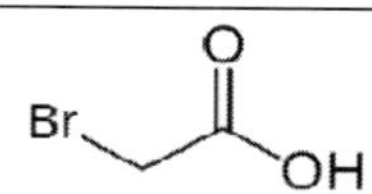	$CH_2BrCOOH$	138.9	2.73
Dichloroacetic acid	DCAA	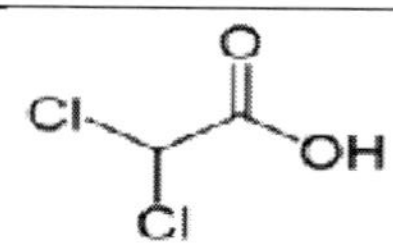	$CHCl_2COOH$	128.9	1.37
Dibromoacetic acid	DBAA	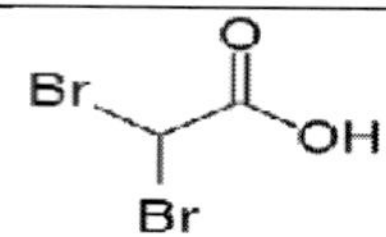	$CHBr_2COOH$	217.8	1.47
Bromochloroacetic acid	BCAA	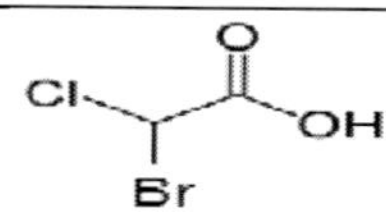	$CHBrClCOOH$	173.4	1.39

Name	Abbreviation	Structure	Abbreviation	Molecular weight ($g \cdot mol^{-1}$)	pK_a
Trichloroacetic acid	TCAA	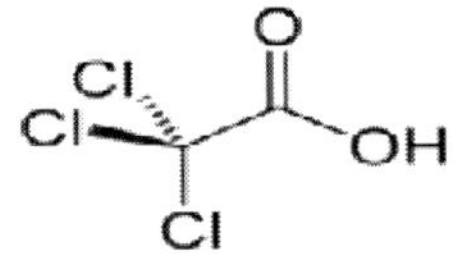	CCl_3COOH	147.7	0.09
Tribromoacetic acid	TBAA	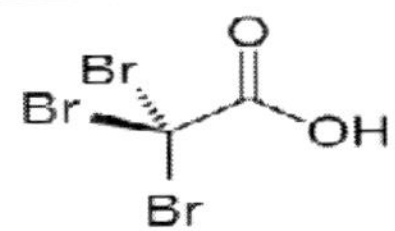	CBr_3COOH	296.7	0.22
Bromodichloroacetic acid	BDCAA	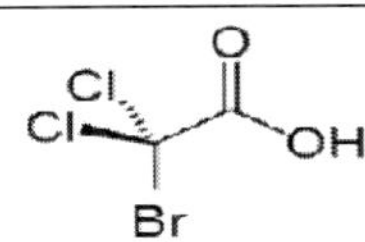	$CBrCl_2COOH$	107.9	0.05
Chlorodibromoacetic acid	CDBAA	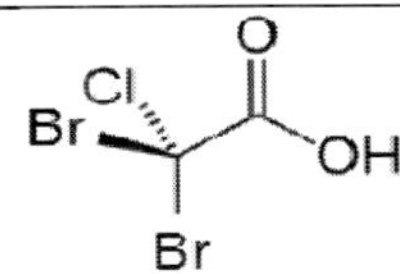	$CBr_2ClCOOH$	252.3	0.13

Recent epidemiological studies have pointed out a possible association between the exposure to HAAs and adverse effects on humans, regarding their implication in colon and liver cancers, as well as effects on the male reproductive system [12, 13].

Humans are exposed to HAAs by different pathways, by drinking water containing HAAs, or by utilizing such waters for cooking, showering, and other similar activities [14].

In addition, waters ingested in swimming pools are expected to be an important route of exposure, because in these waters HAAs can reach high concentrations: between 1 and 800 $\mu g \cdot L^{-1}$ [8, 15]. In drinking waters common levels are lower, but they can reach up to 150 $\mu g \cdot L^{-1}$ [16].

Due to the high toxicity and carcinogenicity of some HAAs and others DBPs [17], the US-EPA has established regulations since 1993. The Stage 1 is the rule of disinfection byproducts (D-DBP I), which established a maximum contaminant level (MCL) of 60 $\mu g \cdot L^{-1}$ for the sum of five haloacetic acids (HAA5s) [18].

The Stage 2 is the rule D-DBP II, which provided a reduction of such MCL to 30 $\mu g \cdot L^{-1}$, in addition to considering BDCA, CDBAA and TBAA.

Table 2 shows some toxicological classification categories for HAAs according to the US-EPA.

Table 2. Toxicological classification for HAAs according to the US-EPA

Compound	US-EPA classification	Adverse effects
Dichloroacetic acid	B2 probable human carcinogen (sufficient laboratory evidence)	Cancer, effects on reproduction and development
Trichloroacetic acid	C possible human carcinogen	Liver, kidneys, spleen, effects on development

In Europe, the Directive 98/83/EC on the quality of waters intended for human consumption did not set maximum levels for haloacetic acids. The World Health Organization (WHO) has established levels of 20 $\mu g \cdot L^{-1}$ for MCAA, 50 $\mu g \cdot L^{-1}$ for DCCA, and 200 $\mu g \cdot L^{-1}$ for TCAA [19].

3. Analytical Methods to Monitor HAAs in Waters

Main current methods to determine HAAs in waters involve gas chromatography (GC), liquid chromatography (LC), and ion chromatography (IC) [20]. Due to the low levels of HAAs in waters, it is advisable the utilization of these chromatographic techniques in combination with mass spectrometric detection.

It is also worthy to mention that HAAs are generally difficult to determine due to their strong acidity and their hydrophilic properties. This, in combination with their low levels in waters, justifies the complexity of the extraction/preconcentration methods required in their determination.

There is not any recent review in the literature covering analytical methods for determining HAAs in water. Most recent work of Paull et al. only focused on the determination of HAAs by IC, and it was published in 2004 [21]. Given the importance of the monitoring of HAAs in waters, due to health implications, it is of utmost relevance to review the most recent works concerning current analytical methods.

Additionally, this book chapter intends to highlight novel microextraction methods that minimize the consumption of toxic organic solvents during the extraction step. Unfortunately, organic solvents are often used in large amounts in conventional methodologies, which is contradictory from an environmental point of view.

3.1. Gas Chromatography to Determine HAAs

Gas chromatography is the most widely used analytical technique for determining HAAs in water, due to several advantages such as high resolution, rapid separation, relatively low cost and easy coupling to sensitive and selective detectors. Table 3 includes most significant literature works which take use of GC to monitor HAAs in waters.

The US-EPA proposes various methods for determining HAAs in drinking waters, in all cases requiring complicated extraction and derivatization steps before GC analysis. Thus, the US-EPA Method 552 [22] and the Method 552.2 [23] uses liquid-liquid extraction (LLE), derivatization, and quantification by GC with electron capture detector (ECD).

Table 3. GC methods to monitor HAAs in waters

HAAs	Matrix	Derivatization reagent	Extraction technique	Detection	LOD ($\mu g \cdot L^{-1}$)	RSD (%)	Ref
9 HAAs	tap & pool waters	DMS[a] and $TBA\text{-}HSO_4$[b]	LLME-HS-SPME	MS	0.33 – 1.5	< 12.7	[32]
5 HAAs	tap & bottled waters	methanol	HF-LPME[c]	ECD	0.1 – 1.8	< 12	[27]
6 HAAs	river waters	n-octanol and TFAA[d]	SDME[e]	MS	0.1 – 1.2	[f]< 12.3	[20]
MCAA	drinking waters	TFE[g]	LLE or SPE	ECD	3	< 10.5	[35]
9 HAAs	drinking waters	methanol	SPE-LLME	MS	0.009 – 0.42	< 12.5	[37]
2 HAAs	drinking waters	methanol	HS	ECD	0.5 and 3	< 21.3	[28]
6 HAAs	tap & bottled waters	methanol	LLE	ECD	0.01 – 0.2	-	[29]
5 HAAs	drinking waters	methanol	LLE-HS-SPME	ECD	0.02 – 0.7	< 12	[30]
10 HAAs	drinking waters	methanol	LLE	MS	0.012 – 0.079	[f]< 6.86	[36]
9 HAAs	pool waters	DMS[a]	HS-SPME	ECD		< 25	[34]
9 HAAs	tap & pool waters	DMS[a] or DES[h] & $TBA\text{-}HSO_4$[b]	HS-SPME	MS	0.01 – 0.45	< 15.6	[33]
9 HAAs	tap & pool waters	DMS[a] and $TBA\text{-}HSO_4$[b]	LLME-HS	MS	0.02 – 0.4	< 11.5	[15]
9 HAAs	drinking waters	PFB-Br[i]	LLE	MS	0.02 – 1	[f]< 19.7	[31]

[a]dimethyl sulfate
[b]tetrabutyl ammonium hydrogen sulfate (ion pair)
[c]hollow-fiber liquid-phase microextraction
[d]trifluoroacetic acid (reaction catalyst)
[e]single-drop microextraction
[f]intraday precision
[g]trifluoro ethanol
[h]diethyl sulfate
[i]pentafluorobenzyl bromide.

The US-EPA Method 552.1 [24] employs solid-liquid extraction (SLE), derivatization, and analysis by GC-ECD. Finally, the US-EPA Method 552.3 [25] uses liquid-liquid microextraction (LLME), followed by derivatization and GC-ECD for measuring the HAA9s. The US-EPA also recognizes the American Public Health Association (APHA) Standard Method 6251B as equivalent to the standard methods of the US-EPA for the analysis of HAA5s in drinking water. As the above mentioned ones, this method includes extraction and derivatization steps, and analysis by GC-ECD.

3.1.1. Derivatization in GC

As abovementioned, all GC methods require a previous step in which HAAs suffer derivatization. HAAs are normally transformed into methylated esters to obtain more volatile and less polar compounds for GC analysis, by utilizing a Fisher esterification reaction, with an alcohol in an acid medium. The most common medium acid in this reaction for HAAs is sulfuric acid. Chloric acid is not used to avoid chlorination.

The US-EPA Method 552 employs LLE with methyl-*tert*-butyl ether (MTBE) as extraction solvent and diazomethane as derivatizing agent. The US-EPA Method 552.1 uses SPE with ion exchange resins, using in this case methanol as derivatizing agent. The US-EPA Method 552.2 uses LLE (with MTBE), and acidic methanol to ensure derivatization. The US-EPA Method 552.3 employs LLME (with previous acidification) using MTBE or *tert*-amyl-methyl ether (TAME), followed by the addition of acidic methanol as derivatizing agent. The majority of US-EPA methods recommend the addition of sodium sulfate and sulfuric acid to the samples in order increase the extraction efficiency.

The esterification processes used to transform HAAs in their respective methylated esters, make these compounds less polar, more stable and volatile. However, the thermal instability of HAAs is an important problem that must be controlled throughout the process since there might be degradation [15]. Thus, HAAs can transform into trihalomethanes by decarboxylation, and also even by the possible hydrolysis of the methylated esters already formed after the derivatization reaction. Indeed, HAAs could also be decomposed into the injection port of a GC at elevated temperatures, especially in the splitless injection mode [26]. Furthermore, it is important to control the water content in the sample when methylated esters are formed, since water can increase the rate of hydrolysis of such esters in the GC injection port, which would also increase the generation of THMs. Particular problems are suffered by tribromoacetic acid (TBAA), whose corresponding ester is the least stable.

Thus, TBAA is degraded to bromoform, and this is believe to be favored if there is some kind of contamination with peroxide. Even low concentrations of peroxides (almost undetectable) can generate significant levels of bromoform in the standard solutions. It is suggested the conservation under nitrogen to minimize the formation of peroxides. In this sense, derivatization is necessary for GC determination, but it needs to be carefully controlled to avoid problems.

This important problem has also been pointed out in US-EPA methods. Thus, US-EPA Method 552.1 does not include TBAA determination. The US-EPA Method 552.2 comments that the acid methylation with methanol leads to a partial decarboxylation of the TBAA and therefore, a peak belonging to bromoform can be observed in the chromatograms. The US-EPA Method 552.3 also comments that brominated trihaloacetic acids can exhibit degradation during storage.

Table 3 includes a list of most common derivatization reagents currently used. Diazomethane was firstly proposed by the US-EPA Method 552 [22]. However, it is highly toxic, unstable, and presents a hazardous preparation. Acidic methanol has been proposed to replace diazomethane for the acid esterification, in combination with GC and ECD [27-30]. This procedure involves the extraction of HAAs from water using MTBE, followed by esterification in the organic phase. Other derivatizing agents have been proposed, such as pentafluorobenzyl bromide (PFBBr) [31], dimethyl sulfate (DMS) [15, 32-34], diethyl sulfate (DES) [33], trifluoroethanol [35], and n-octanol, using trifluoroacetic anhydride as catalyst in this particular case [35].

It is also helpful the formation of ion pairs for HAAs to favor their extraction and derivatization, and with this purpose quaternary ammonium salts like tetrabutylammonium bromide or tetrabutylammonium chloride have been used [15, 32, 33].

3.1.2. Extraction Techniques Combined with GC

A number of extraction techniques, in combination with derivatization, are used prior to GC. Despite the relative large amount of organic solvents required to perform LLE, this extraction technique is still widely used. Thus, Minghong et al. [31] used LLE-GC-MS with MTBE as extraction solvent, and pentafluorobenzyl bromide as derivatizating agent. Leivadara et al. proposed a similar LLE approach as described in US-EPA 552, employing MTBE as extractant solvent, and acidic methanol as derivatizing reagent for six HAAs, in conjunction with GC and ECD [29]. In 2002, Pervova et al. proposed a LLE method which uses diethyl ether as extraction solvent and TFE as derivatizing agent, being the rest of the method similar to US-EPA 552. Authors also compare the performance of the method with that of SPE for monitoring MCAA in drinking waters [35]. In a recent study, Wei et al. [36], determined 10 HAAs by LLE using MTBE as extraction solvent and acidic methanol as derivatization reagent. In this case, the analysis was performed using GC and MS/MS.

Several extraction techniques combined with GC are based on the determination of derivatized HAAs in their gas phase. Thus, headspace (HS-) approaches are very common. The basis of the method is the distribution of the volatile halogenated organic compounds between the aqueous sample and the gas phase. The simplest approach utilizes headspace sampling (manual or automatic) with a syringe. Thus, Wong et al. [28] have used HS-GC-ECD, in this case for determining DCAA and TCAA, employing a previous esterification with acidic methanol. It must be noted that the HS- procedure is normally accomplished after a previous LLE or LLME extraction and derivatization (LLE or LLME-derivatization-HS-GC). As an example, Gallego et al. [15] used a rapid step of LLME with a very polar solvent (n-pentane), and simultaneous derivatization with dimethyl sulfate (DMS) with the aid of tetrabutylammonium salt to form ion pairs, followed by co-evaporation of methylated esters and HS-GC-MS determination. Authors also observed thermal decomposition of the methylated esters at different injection temperatures: and so they decide to utilize 250 ºC [15].

HS in combination with solid-phase microextraction (SPME) has also been proposed. The first work belongs to Sarrion et al. which perform an *in situ* derivatization with DMS or DES (adding TBA-HSO_4 as ion pair reagent), followed by HS-SPME-GC-MS with the commercial fiber carboxen-polydimethylsiloxane (CAR-PDMS) [33]. Sá et al. [34] also determined 9 HAAs after derivatization to their respective methylated esters with DMS, using the commercial fiber CAR-PDMS and HS-SPME-GC-ECD. Some authors have also combined the utilization of LLE with HS-SPME. Thus, Hammami et al. [30] performed a previous LLE method (similar to US-EPA Method 552) using MTBE as extractant solvent, which was further evaporated. The extract was then reconstituted with acidic methanol to perform derivatization for 10 min. Afterwards, the vial is heated and derivatized analytes are extracted by HS-SPME using the commercial fiber polydimethylsiloxane (PDMS) for another 10 min. Cardador et al. [32] have also carried out a LLME-HS-SPME, using n-pentane as extraction solvent, and derivatization with DMS (with the aid of TBA-HSO_4 as ion pair reagent). In this case, CAR-PDMS was used as the commercial SPME fiber.

Solid-phase extraction (SPE) is another successful extraction and preconcentration method, based on the use of trapping sorbents. Sadia et al. [37] determined 9 HAAs in drinking waters by SPE with simultaneous elution and derivatization with acidic methanol, followed by a post LLME with MTBE and GC-MS. Pervova et al. [35] also employed SPE, but in this case elution and derivatization did not occur simultaneously. Authors eluted HAAs with diethyl ether, and then the derivatization was carried out with trifluoroethanol in hexane to obtain the trifluoroacetic esters. Furthermore, this technique was compared with LLE, also using diethyl ether as the extraction solvent and trifluoroethanol for esterification. In this case, higher recoveries and performance were obtained with SPE.

Within the trend of microextraction approaches, the determination of HAAs has also been carried out by different liquid-phase microextraction (LPME) methods.

Thus, Varanusupakul et al. [27] used a hollow-fiber liquid phase microextraction (HF-LPME) method, also named as supported liquid-liquid microextraction (SLLME), in this particular case in HS- mode. The technique uses few microliters of an extractant solvent, which is placed on a hollow fiber (HF) or membrane to increase its stability, and supported onto the headspace of the extraction vial.

Derivatization was conducted with acidic methanol, and extraction of the methylated esters was carried out in with the SLLME (containing 20 μL of octanol supported onto the HF) located in the HS. The overall extraction was 60 minutes, and the method was used together with GC-ECD. Another mode of LPME is single-drop microextraction (SDME), which uses a droplet (also few microliters) of and extractant solvent directly suspended at the tip of a syringe to carry out the extraction method. It has been utilized for HAAs in combination with GC-MS by Saraji et al. [20], using n-octanol (1.8 μL) as both the extraction solvent and the derivatizing agent. In this procedure, trifluoroacetic anhydride (TFAA) was used as the reaction catalyst, needing 20 min for the overall extraction/derivatization procedure. N,O-bis(trimethylsilyl) acetamide (bisTMSA) and PFBBr were also tested as derivatizing agents.

All GC methods for the determination of HAAs are characterized by low detection limits, ranging from 0.009 [37] to 0.45 $\mu g \cdot L^{-1}$ [33], and adequate precision performance, with relative standard deviation (RSD) values (in %), always lower than 19.7%.

3.2. Liquid Chromatography to Determine HAAs

Liquid chromatography is a useful technique for the determination of HAAs in aqueous samples, with advantages over GC since it does not require a previous derivatization step. Its advantages are even higher if mass or tandem mass spectrometry (LC–MS or LC–MS/MS) detection is used, due to its sensitivity and specificity [38-40].

LC conditions, especially mobile phase composition and additives, extremely influence the separation and sensitivity of the HAAs, given their high acidic nature. In addition, the appropriate selection of chromatographic column is also important given the inherent polar characteristics of HAAs [40, 41]. Reversed-phase liquid chromatography (RP–LC) is the most commonly used type of LC; however, the non-polar stationary phase generally has difficulty to retain small, charged polar molecules such as HAAs [42]. Thus, ion chromatography (IC) and also hydrophilic interaction liquid chromatography (HILIC) [40] are normally preferred.

Table 4 includes most significant literature works regarding LC for the monitoring of HAAs in waters, excluding IC applications (which will be later described).

It is important to mention that several authors have carried out the determination of HAAs without the need of any extraction/preconcentration method, mainly taken advantage of the MS/MS detection. Thus, Meng et al. [43] developed a sensitive method to determine 9 HAAs in drinking waters using LC–MS/MS. Authors utilized a C8 RP–LC column, rather than the common C18, and without the use of ion-pairing reagents. Mobile phase composed of acetonitrile and ultrapure water containing 0.1% (v/v) acetic acid (pH value of 3.1) was used under gradient conditions, obtaining the separation of HAAs in 7.5 min.

Detection limits ranged from 0.16 to 0.99 $\mu g \cdot L^{-1}$, being the exceptions CDBAA with 1.44 $\mu g \cdot L^{-1}$ and TBAA with 8.87 $\mu g \cdot L^{-1}$. Duan et al. [38] developed a rapid method for 9 HAAs using ultra-high-performance liquid chromatography–electrospray ionization–tandem mass spectrometry (UHPLC–ESI–MS/MS). Authors utilized a RP C18 column but with a certain hydrophilic character developed for detection of polar analytes, specifically HSS T3. Separation was accomplished in gradient mode with a mobile phase composed of methanol and deionized water with 0.125 $mmol \cdot L^{-1}$ in formic acid. The limits of detection ranged between 0.06 and 0.16 $\mu g \cdot L^{-1}$ without the need of any preconcentration step.

Table 4. LC methods (excluding IC) to monitor HAAs in waters

HAAs	Matrix	Extraction method	LC requirements	Detection	LOD ($\mu g \cdot L^{-1}$)	E_F[a] / E_R[b] (%)	RSD (%)	Ref.
9 HAAs	drinking waters	-	C8 column	MS/MS	0.16 – 8.87	- / 80.1 – 108	[c]< 12	[43]
9 HAAs	drinking, surface & ground waters	-	HSS T3 column (UHPLC)	ESI–MS/MS	0.06 – 0.16	- / 88 – 110	[c]< 6.1	[38]
9 HAAs	waste waters	SPE[d] to remove DOM[e]	HSS T3 column (UHPLC)	ESI–MS/MS	0.08 – 0.25	- /91.9 – 104.2	< 8.2	[39]
9 HAAs	tap waters	SLMME[f] with DHE[g] and back-extraction with NaOH	C18 column	UV	0.02 – 2.69	324 – 3298 / 3.89 – 39.6	< 10.8	[44]
9 HAAs	waste waters	SLMME[f] with DHE[g] and back-extraction with NaOH	C18 column	UV	0.1 – 6.84	71.3 – 428.4 / 8.9 – 53.6	[c]< 10.3	[45]
9 HAAs	waste waters	SLMME[f] with DHE[g] and back-extraction with mobile phase	C18 column	UV	2.23 – 107	7.1 – 14.9 / 26.4 – 34.1	< 8.54	[46]
2 HAAs	waste waters	EME[h] with SLM[i] comparison with SPE	C18 column	UV	0.72 and 19.3 $ng \cdot L^{-1}$	- / 87 – 106	< 4.9	[47]

HAAs	Matrix	Extraction method	LC requirements	Detection	LOD ($\mu g \cdot L^{-1}$)	E_F[a] / E_R[b] (%)	RSD (%)	Ref.
9 HAAs	drinking waters	vaporization under vacuum	BetaMax acid column (UHPLC)	ESI–MS/MS	0.18 – 71.5	40 – 400/ 69.7–114	[c]< 27.8	[40]
9 HAAs	swimming pool waters	μ–SPE[j]	C18 column (UHPLC)	UV	0.001 – 0.092	- / <110	< 7.4	[48]

[a]enrichment / preconcentration factor
[b]extraction efficiency
[c]intraday precision
[d]solid-phase extraction
[e]dissolved organic matter
[f]supported liquid membrane microextraction
[g]dihexyl ether
[h]electro membrane extraction
[i]supported liquid membrane
[j]micro solid phase extraction.

These authors [39] have also reported a similar method for the determination of 9 HAAs in wastewater effluents by UHPLC–ESI–MS/MS (same LC column and mobile phase) in the multiple reaction monitoring (MRM) mode. The method required a prior elimination of the dissolved organic matter (DOM) from effluent samples by utilizing solid-phase extraction. In this case, the limits of detection ranged between 0.08 and 0.25 $\mu g \cdot L^{-1}$, which should be highlighted given the complexity of the wastewaters.

3.2.1. Extraction Techniques Combined with LC

Different membrane extraction modes have been successfully applied to the determination of HAAs in waters. The most employed technique for sample pretreatment is membrane extraction, specifically supported liquid membrane microextraction (SLMME).

Wang et al. [44, 46] have developed various determination methods for HAAs with HPLC–UV employing a supported liquid membrane (SLM) as alternative extraction method. Thus, authors utilized a polypropylene membrane which served as support of the extraction solvent (few microliters), for aqueous samples which were acidified with sulfuric acid. The preferred extraction solvent was dihexyl ether. The extraction time depends on the extraction mode, thus, shorter times of 15 and 30 min were needed in the more generic SLLME mode [45, 46], whereas up to 60 min are required if SLLME is used as hollow-fiber liquid-liquid microextraction (HF–LLME) [44]. After extraction, authors performed a back-extraction with NaOH [44, 45] or with the LC mobile phase [46] to ensure compatibility of the HAAs extracted with the chromatographic system. Authors obtained LODs down to 107 $\mu g \cdot L^{-1}$ [46]. Enrichment factors between 65 and 500 were obtained, depending on the SLLME mode [44, 46]. In all works, the HPLC mobile phase was 15 mM KH_2PO_4 (pH value of 2.2): acetonitrile 95:5 (v/v).

Alhooshani et al. [47] determined 2 HAAs in wastewater samples using electromembrane extraction (EME) in combination with supported liquid membrane (SLM) followed by HPLC–UV. The obtained LODs were 0.072 $ng \cdot L^{-1}$ for MCAA and 19.3 $ng \cdot L^{-1}$ for DCAA. Authors even compare the performance of their method with that of SPE, getting higher recoveries.

Chen et al. [40] have determined 9 HAAs in drinking waters by UHPLC–MS/MS. Authors performed a preconcentration step of samples by vaporization under vacuum. MS detection was performed in negative ESI-MS/MS. Authors compared the performance of the BetaMax Acid column with a conventional HILIC column. Despite the fact that chromatographic detection limits were lower when a conventional UHPLC HILIC column was used (0.08–2.73 $\mu g \cdot L^{-1}$), with respect to those obtained when using the BetaMax Acid column (0.18–71.5 $\mu g \cdot L^{-1}$), authors showed that it was more suitable the BetaMax Acid column. The reason was that the BetaMax Acid column did not require any sample preparation, whereas the conventional HILIC UHPLC column required dilution of water samples in 90% (v/v) of acetonitrile.

Micro solid–phase extraction (μ–SPE) has also been employed for the determination of HAAs. Thus, Nsubuga et al. [48] determined 9 HAAs in swimming pool waters by membrane-protected μ–SPE and UHPLC-UV. A novel sorbent was developed using rice husk. Iron oxide was incorporated via a sol-gel process to increase the extraction capability of rice husk silica. The obtained LODs were really low despite UV detection, ranging between 0.001 and 0.092 $\mu g \cdot L^{-1}$.

In any case, all LC methods reported for the determination of HAAs fulfill quality analytical requirements, with adequate precision performance having relative standard deviation (RSD) values (in %) always lower than 27.8% [40].

3.3. Ion Chromatography to Determine HAAs

Ion chromatography (IC) possesses an enormous applicability for the determination of HAAs, and thus it deserves a whole section by itself in spite of being a liquid chromatographic mode. The highly polar character of HAAs and their hydrophilic properties justify the fact that these analytes become adequate to ionic exchange [49]. HAAs are moderately strong acids, and so they are commonly dissociated to their corresponding haloacetates in typical drinking water conditions (pH < 6). Table 5 includes a summary of recent works of literature which used IC for determining HAAs in waters.

The US-EPA Method 557 has been proposed for the determination of haloacetic acids, bromate, and dalapon in drinking waters by IC electrospray ionization tandem mass spectrometry (IC-ESI-MS/MS) [50]. The method utilizes ammonium chloride to react with residual chlorine present in drinking water samples to form chloramines, effectively preventing chlorine-mediated formation of analytes during storage. The non-treated sample is then directly injected onto an IC column specifically designed to separate HAAs from the following common anions in drinking waters: chloride, carbonate, sulfate, and nitrate.

Table 5. IC methods to monitor HAAs in waters

HAAs	Matrix	Extraction technique	Ion exchange column /suppressor	Mobile phase	Detection	LOD ($\mu g \cdot L^{-1}$)	RSD (%)	Ref.
9 HAAs	drinking water	-	IonPac AS9-SC /-	NaOH & H_3BO_3	FD[a]	0.6 – 10.1	< 32	[60]
9 HAAs	tap water	SPE -cartridge: LiChrolut EN -clean-up: IC-Ba, IC-Ag, IC-H	IonPac Cryptand A1/ASRS	NaOH & LiOH	ESI-MS	1.5 – 10	< 11	[62]
5 HAAs	drinking water	-	IonPac AS9-SC /-	NaOH & H_3BO_3	FD[a]	0.6 – 2.9	< 21	[61]
9 HAAs	drinking water	-	IonPac AS16 or IonPac AS9-HC / ASRS and Atlas	NaOH & Na_2CO_3	CD	1.11 – 9.32[b] 0.37–31.64[c]	< 8.1[b] < 3.5[c]	[52]
9 HAAs	drinking water	-	IonPac AS16 / ASRS	NaOH	ICP-MS	0.34 – 0.99	< 4.95	[49]
7 HAAs	drinking water	SPE -cartridge: LiChrolut EN -clean-up: IC-Ba, IC-Ag, IC-H	IonPac AS16 / ASRS and Atlas	KOH	CD	0.09 – 21.5	< 4.9	[53]
7 HAAs	drinking water	SPE -cartridge: LiChrolut EN -clean-up: IC-Ba, IC-Ag, IC-H	IonPac AS16 / Atlas	NaOH	CD ESI-S/MS	0.02 – 2.46[d] 2 – 138[e]	< 9.2[d] < 13[e]	[54]
6 HAAs	drinking, river and pool waters	-	IonPac AS11-SC /-	NH_4NO_3	ICP-MS	1.36 – 3.28	< 12.8	[58]

HAAs	Matrix	Extraction technique	Ion exchange column /suppressor	Mobile phase	Detection	LOD ($\mu g \cdot L^{-1}$)	RSD (%)	Ref.
3 HAAs	bottle water	MAE[f]	IonPac AS9-HC / Atlas	Na_2CO_3	CD	0.06 – 0.85	< 2.71	[59]
9 HAAs	tap, bottle and waste waters	SPE -cartridge: LiChrolut EN -clean-up: IC-Ba, IC-Ag	IonPac AS11-HC / ASRS ultra	KOH	CD	0.7 – 30	< 7.62	[57]
3 HAAs	drinking water	SPE -cartridge: LiChrolut EN -clean-up: IC-acid	Shodex IC SI 90 4E /AMMS III	MeOH & Na_2CO_3	CD	0.05 – 0.36	< 2.3	[63]
9 HAAs	drinking water	SPE -cartridge: LiChrolut EN -clean-up: IC-Ba, IC-Ag, IC-H	IonPac AS16 /Atlas -temp. programming to improve resolution	NaOH	CD	-	-	[55]
5 HAAs	drinking water	SPE -cartridge: LiChrolut EN -clean-up: IC-Ba, IC-Ag, IC-H	IonPac AS16 or IonPac AS11-HC /ASRS	KOH	CD	130 – 472.6[b] 1.89 – 11.8[g]	< 3.0[b] < 7.9[g]	[56]
7 HAAs	drinking water	SPE (previous cleaning: C18) -cartridge: LiChrolut EN -clean-up: IC-Ba, IC-Ag, IC-H	IonPac AS16 / ASRS	NaOH	CD	0.48 – 2.61	< 3	[51]

[a]post-columnderivatization with nicotylamide
[b]results using IonPac AS16 column
[c]results using IonPac AS9HC column
[d]results obtaining with CD
[e]results obtaining with ESI-MS
[f]microwaves-assisted extraction
[g]results using IonPac AS11-HC column.

The matrix components in the column eluate are then monitored via conductivity detection (CD) and then diverted to waste; whereas HAAs are directed into the ESI-MS/MS system. Acetonitrile is added post-column to enhance desolvation of HAAs in the ESI interface.

One of the most important parameters in IC is the selection of the analytical column, because it must perform with specific features given the retention of these polar analytes. Among them, it can be cited: IonPac AS16 [49, 51-56]; IonPac AS11-HC [56-58]; IonPac AS9-HC [52, 59]; IonPac AS9-SC [60, 61]; IonPac Cryptand A1 [62]; and Shodex IC SI 90 4E column [63]. It is important to mention here that there is an ion exchange column with high capacity, IonPac AS24, which has been specifically designed for HAAs analysis [64], being particularly adequate for US-EPA method 557. This column has a similar chemistry to that of AS19-AS20, and it is recommended for MS detection and matrices with high ionic strengths which are subjected to analysis without any prior pretreatment.

Depending on the detector used in combination with IC, it is normally necessary to use a suppressor column mainly to decrease the conductivity of the mobile phase used in the elution. The most common suppressor columns used for the determination of HAAs are ASRS [49, 51-53, 56, 57, 62] and Atlas [52-55, 59]. A post-column derivatization reaction is needed when fluorescence detection (FD) is used [60, 61].

3.3.1. Extraction Techniques Combined with IC

Several works do not use any sample treatment neither sample extraction/preconcentration to determine HAAs in waters when using IC [49, 58, 60, 61]. This highly depends on the detection system used and/or the sensitivity required. Thus, if inductively coupled plasma (ICP) in conjunction with MS is used, there is no need of sample extraction/preconcentration [48, 58]. The same situation appears if FD is employed [60, 61]. If an adequate MS/MS is carried out, like in US-EPA method 557, it is also possible to avoid sample preconcentration. If FD is used, detection limits down to 10.1 $\mu g \cdot L^{-1}$ [60] can be obtained. If ICP-MS is used, detection limits down to 3.28 $\mu g \cdot L^{-1}$ [58] have also been described.

In any case, the most common sample preconcentration method prior to IC is SPE [51, 53-57, 62] to reach adequate sensitivity performance. It must be considered that if SPE is used, water needs to be acidified before preconcentration and the cartridge needs pretreatment to remove chlorine and sulphate ions. These steps are necessary to avoid interferences and so to improve IC sensitivity and selectivity, due to the presence of these ions in

$mg \cdot L^{-1}$ levels in the water matrix [21]. As pointed out by Paull and Barron the elimination of these anions from the concentrated sample can be done using cation-exchange SPE cartridges in the Ag form (to eliminate the excess of chlorine) followed by a cation-exchange SPE cartridge in the acid form (H^+) to remove Ag ions afterwards [51, 53-57, 62], or in the Ba form (to remove the excess of sulphate) [51, 53-57, 62]. Indeed, the majority of works use both post-columns clean ups. Among SPE brands, most commonly used cartridge is LiChrolut EN [51, 53-57, 62]. The obtained preconcentration factors with SPE for HAAs varied from 20 [57, 61] to 25 [52-54, 56, 62], with detection limits varying from 0.02 $\mu g \cdot L^{-1}$ [54] to 472.6 $\mu g \cdot L^{-1}$ [56] depending on the detector utilized.

In 2004, Liu et al. also proposed the use of a microwaves-assisted extraction [59] rather than SPE, obtaining recoveries between 91 and 93% for MCAA, DCAA and TCAA, and limits of detection between 0.06 and 0.85 $\mu g \cdot L^{-1}$.

3.4. Other Techniques to Determine HAAs

3.4.1. Capillary Electrophoresis

Capillary electrophoresis is another technique that has been described for the determination of HAAs in water. As ion chromatography, it takes advantage of the intrinsic ionic properties of HAAs.

Focusing on recent works and trends, Ding et al. in 2010 [65] employed CE with contactless conductivity detector in a microchip, in combination with SPE. Authors performed a previous conventional SPE clean-up step, followed by SPE on a chip. The combination of the microchip with the contactless conductivity detector shows the applicability of the "lab-on-a-chip" system for the screening of HAAs in waters. Authors obtained LODs between 38 and 62 $\mu g \cdot L^{-1}$ for TCAA and DCAA, respectively.

Zhang et al. in 2011 used CE with pressure-assisted electrokinetic injection, and electrospray ionization - tandem mass spectrometry as detector (PAEKI-ESI-MS) [66]. In this case, it was not necessary to reverse the polarity of the CE to achieve the separation. The pressure of the EKI was varied to improve the elution of HAAs, getting preconcentration factors up to 20000 for 9 HAAs plus monoiodoacetic acid (MIAA).

Kubáň et al. in 2012 utilized CE with a contactless conductivity detector to determine 5 HAAs in waters altogether with other analytes. The procedure was used with SPE as extraction technique. The overall method was characterized for achieving limits of detection between 6.1 and 12.2 $\mu g \cdot L^{-1}$ for the studied HAAs [67].

3.4.2. Sensors

It is clear the interest of developing sensors in environmental monitoring due to their fast response accompanied by high selectivity and sensitivity [68].

The majority of the works which use sensors for the determination of HAAs are composed of molecularly imprinted polymers (MIPs) [69-72]. Thus, Suedee et al. [71] used a MIP imprinted with TCAA via two procedures: bulk polymerization (BP) and multi-step swelling polymerization (MSP). Authors determined 6 HAAs in waters achieving LODs down to 5 $\mu g \cdot L^{-1}$. The US-EPA Method 552 was used as a confirmatory method regarding the presence of HAAs in waters. In 2006, the same group modified the polymer membrane by adding gold particles that performed as the electrode. The overall method was able to determine 6 HAAs in only 2 min [70]. Later that year, authors used a piezoelectric quartz crystal microbalance (QCM) with a multi-electrode. The coated QCM was able to specifically detect 6 HAAs in water samples in terms of the mass change in relation to acid–base interactions of the analytes with the MIP [72]. The last work of Suedee in this field was reported in 2007, also using the QCM technique, but having the piezoelectric material imprinted with TCAA and polypyrrole [69]. Thus, author obtained LODs between 35 and 50 $\mu g \cdot L^{-1}$ for the studied HAAs.

3.4.3. Flow Injection Analysis (FIA)

In 2005, Geme et al. have developed a capillary membrane sampling-flow injection analysis (CMS-FIA) for determining 5 HAAs and 4 THMs [73]. The procedure is based on the derivatization reaction of HAAs with nicotinamide to obtain fluorescence compounds. However, the CMS-FIA method is only able to obtain the total content of 5 HAAs. The overall system is able to analyze 2-3 samples per hour obtaining LODs lower than 1.6 $\mu g \cdot L^{-1}$. The same authors have applied this technique in water distribution systems [74]. They develop a simple analytical instrument to directly carry out measurements of up to 9 HAAs (as total content) in near-real-time.

3.4.4. Improvements in Mass Spectrometry Determinations of HAAs

Several authors have described improvements in MS detection without the need of a previous chromatographic or electrophoretic separation. Thus, Zhai et al. [75] determined 9 HAAs by ESI-MS using a previous step of water extraction with LLE using MTBE and high ionic strength. Afterwards, MTBE is evaporated down to 0.5 mL, reconstituted with acetonitrile, and subjected to ESI-MS. Authors were able to differentiate in MS those adducts/dimers coming from HAAs than those of higher molecular weights using the analysis mode: in-electrospray ionization–mass spectrometry. Zhang et al. [76] have developed a method to carry out a selective detection of brominated DBP or highly polar DBPs using negative ion electrospray ionization–triple quadrupole mass spectrometry, after a previous step of LLE using MTBE as extraction solvent.

ACKNOWLEDGMENTS

This work has been funded by the project ProID20100241 of the Agencia Canaria de Investigación, Innovación y Sociedad de la Información (ACIISI) under the European Fund for Regional Development (FEDER), and by Project Ref. AGUA05-SPDs from Fundación CajaCanarias. V. Pino thanks the Spanish Ministry of Economy and Competitiveness (MINECO) for the Ramón y Cajal contract with the University of La Laguna (ULL). P. Rocío-Bautista thanks the Fundación Caja Canarias Fellowship with ULL.

REFERENCES

[1] Rook, J. J. *Water Treat. Exam.* 1974, 23, 234–243.
[2] Christman, R. F.; Norwood, D. L.; Millington, D. S.; Johnson, J. D.; Stevens, A. A. *Environ. Sci. Technol.* 1983, 17, 625–628.
[3] Miller, J. W.; Uden, P. C. *Environ. Sci. Technol.* 1983, 17, 150–157.
[4] Shi, H.; Qiang, Z.; Adams, C. *Chemosphere*. 2013, 90, 2485–2492.
[5] Tian, C.; Liu, R.; Guo, T.; Liu, H.; Luo, Q.; Qu, J. *Sep. Purif. Technol.* 2012, 102, 86–93.
[6] Richardson, S. D.; Plewa, M. J.; Wagner, E. D.; Schoeny, R.; DeMarini, D. M. *Mutat. Res.* 2007, 636, 178–242.
[7] Hunter III, S. E.; Blanton, M. R.; Rogers, E. H.; Mole, M. L.; Andrews, J.; Chernoff, N. *Reprod. Toxicol.* 2006, 22, 443–448.
[8] Agents Classified by the IARC Monographs, Volumes 1–109 http://monographs.iarc.fr/ENG/Classification/ClassificationsAlphaOrder.pdf (accessed Dec. 2013).

[9] Tao, L.; Wang, W.; Li, L.; Kramer, P. M.; Pereira, M. A. *Toxicol. Sci.* 2004, 82, 62–69.

[10] Melnick, R. L.; Nyska, A.; Foster, P. M.; Roycrof, J. H.; Kissling, G. E. *Toxicology.* 2007, 230, 126–136.

[11] Andrews, J. E.; Nichols, H. P.; Schmid, J. E.; Mole, L. M.; Hunter III, E. S.; Klinefelter, G. R. *Reprod. Toxicol.* 2004, 19, 111–116.

[12] Legay, C.; Rodriguez, M. J.; Sadiq, R.; Sérodes, J. B.; Levallois, P.; Proulx, F. *J. Environ. Manage.*2011, 92, 892–901.

[13] Chowdhury, S.; Rodríguez, M. J.; Sadiq, R. *J. Hazard. Mater.* 2011, 187, 574–584.

[14] Caro, J.; Serrano, A.; Gallego, M. *J. Chromatogr. B.* 2007, 848, 277–282.

[15] Cardador, M. J.; Serrano, A.; Gallego, M. *J. Chromatogr. A.* 2008, 1209, 61–69.

[16] Sánchez-Zafra, A. *Higiene y Sanidad Ambiental.* 2008, 8, 280–290.

[17] Rodríguez, M. J.; Rodríguez, G.; Serones, J.; Sadiq, R. *Interciencia.* 2007, 32, 749–756.

[18] US-EPA (1998). National primary drinking water regulations: disinfectants and disinfection byproducts. Rules and regulation.Vol. 63. http://water.epa.gov/lawsregs/rulesregs/sdwa/mdbp/index.cfm (accessed Dec. 2013).

[19] World Health Organization (WHO) Richardson, S. D.; Postigo, C. (2011). http://whqlibdoc.who.int/publications/2011/9789241548151_eng.pdf. (accessed Dec. 2013).

[20] Saraji, M.; Bidgoli, A. A. H. *J. Chromatogr. A.* 2009, 1216, 1059–1066.

[21] Paull, B.; Barron, L. *J. Chromatogr. A.* 2004, 1046, 1–9.

[22] US-EPA METHOD 552: Hodgeson, J. W.; Collins, J.; Barth, R. E. (1990). Determination of haloacetic acids in drinking water by liquid-liquid extraction, derivatization, and gas chromatography with electron capture detection. http://www.caslab.com/EPA-Method-552 (accessed Dec. 2013).

[23] US-EPA METHOD 552.2 revision 1.0: Munch, D. J.; W. Munch, W.; Pawlecki-Vonderheide, A. M. (1995). Determination of haloacetic acids and dalapon in drinking water by liquid-liquid extraction, derivatization and gas chromatography with electron capture detection. http://dec.alaska.gov/eh/docs/lab/dw/Chem/organics/552.2.pdf (accessed Dec. 2013).

[24] US-EPA METHOD 552.1 revision 1.0: Hodgeson, J. W.; Becker, D. (1992). Determination of haloacetic acids and dalapon in drinking water

by ion-exchange liquid-solid extraction and gas chromatography with an electron capture detector. http://dec.alaska.gov/eh/docs/lab/dw/Chem/organics/552.1.pdf (accessed Dec. 2013).
[25] USEPA METHOD 552.3 revision 1.0: Domino, M. M.; Pepich, B. V.; Munch, D. J.; Fair, P. S.; Xie, Y. (2003). Determination of haloacetic acids and dalapon in drinking water by liquid-liquid microextraction, derivatization, and gas chromatography with electron capture detection. http://www.epa.gov/ogwdw/methods/pdfs/methods/met552_3.pdf (accessed Dec. 2013).
[26] Ma, Y.; Chiang, C. *J.* Chromatogr. A. 2006, 1076, 216–219.
[27] Varanusupakul, P.; Vora-adisak, N.; Pulpoka, B. *Anal.Chim. Acta*. 2007, 598, 82–86.
[28] Wanga, Y. H.; Wong, P. K. *Water Res.* 2005, 39, 1844–1848.
[29] Leivadara, S. V.; Nikolaou, A. D.; Lekkas, T. D. *Food Chem.* 2008, 108, 277–286.
[30] Hammami, B.; Driss, M. R. J. *Anal. Chem.* 2013, 68, 671–679.
[31] Jia, M.; Wu, W. W.; Yost, R. A.; Chadik, P. A.; Stacpoole, P. W.; Henderson, G. H. *Anal. Chem.* 2003, 75, 4065–4080.
[32] Cardador, M. J.; Gallego, M. *Anal. Bioanal. Chem.* 2010, 396, 1331–134.
[33] Sarrión, M. N.; Santos, F. J.; Galceran, M. T. *Anal. Chem.* 2000, 72, 4865–4873.
[34] Sá, C. S. A.; Boaventura, R. A. R.; Pereira, I. B. *J. Environ. Sci. Health Part A-Toxic/Hazard. Subst. Environ. Eng.* 2012, 47, 176–183.
[35] Pervova, M. G.; Kirichenko, V. E.; Pashkevich, K. I. *J. Anal. Chem.* 2002, 57, 326–330.
[36] Li, W.; Liu, Y.; Duan, J.; Mulcahy, D. *Anal. Methods.*2013, 5, 2258–2266.
[37] Sadia, W.; Pauzi, A. *Chromatographia.* 2009, 69, 1447–1451.
[38] Duan, J.; Li, W.; Sia, J.; Mulcahy, D. *Anal. Methods.* 2011, 3, 1667–1673.
[39] Duan, J.; Li, W.; Sun, P.; Lai, Q.; Mulcahy, D.; Guo, S. *Anal. Lett.* 2013, 46, 569–588.
[40] Chen, C. Y.; Chang, S. N.; Wang, G. S. *J. Chromatogr. Sci.* 2009, 47, 67–74.
[41] Dixon, A. M.; Delinskya, D. C.; Bruckner, J. V.; Fisher, J. W.; Bartlett, M. G. *J. Liq. Chromatogr. Relat. Technol.* 2004, 27, 2343–2355.
[42] Delinsky, A. D.; Bruckner, J. V.; Bartlett, M. G. *Biomed. Chromatogr.* 2005, 19, 617–639.

[43] Meng, L.; Wu, S.; Ma, F.; Jia, A.; Hu, J. *J. Chromatogr. A*. 2010, 1217, 4873–4876.
[44] Kou, D.; Wang, X.; Mitra, S. *J. Chromatogr. A*. 2004, 1055, 63–69.
[45] Wang, X.; Kou, D.; Mitra, S. *J. Chromatogr. A*. 2005, 1089, 39–44.
[46] Wang, X.; Saridara, C.; Mitra, S. *Anal. Chim. Acta*. 2005, 543, 92–98.
[47] Alhooshania, K.; Basheera, C.; Kaurb, J.; Gjelstadc, A.; Rasmussenc, K. E.; Pedersen-Bjergaardc, S.; Le, H. K. *Talanta*. 2001, 86, 109–113.
[48] Nsubuga, H.; Basheer, C. *J. Chromatogr. A*. 2013, 1315, 47–52.
[49] Liu, Y.; Mou, S.; Chen, D. *J. Chromatogr. A*. 2004, 1039, 89–95.
[50] US-EPA METHOD 557 revision 1.0: Zaffiro, A. D.; Zimmerman, M.; Pepich, B. V.; Slingsby, R. W.; Jack, R. F.; Pohl, C. A.; Munch, D. J. (2009). Determination of haloacetic acids, bromate, and dalapon in drinking water by ion chromatography electrospray ionization tandem mass spectrometry. http://water.epa.gov/scitech/drinkingwater/labcert/upload/met557.pdf (accessed Dec. 2013).
[51] Ying-xue, S. U. N.; Ping, G. U. J. *Environ. Sci.* 2007, 19, 885–891.
[52] Liu, Y.; Mou, S. *J. Chromatogr. A*. 2003, 997, 225–235.
[53] Barron, L.; Paull, B. *J. Chromatogr. A*. 2004, 1047, 205–212.
[54] Barron, L.; Paull, B. *Talanta*. 2006, 69, 621–630.
[55] Barron, L.; Nesterenko, P. N.; Paull, B. *J. Chromatogr. A*. 2005, 1072, 207–215.
[56] Barron, L.; Paull, B. *Anal. Chim. Acta*. 2004, 522, 153–161.
[57] Uansiri, S.; Kanchanamayoon, W. *J. Agric. Environ. Sci.* 2009, 6, 564–566.
[58] Shi, H.; Adams, C. *Talanta*. 2009, 79, 523–527.
[59] Liu, Y.; Mou, S. *Chemosphere*. 2004, 55, 1253–1258.
[60] Simone, P. S.; Ranaivo, P. L.; Geme, G.; Brown, M. A.; Emmert, G. L. *Anal. Chim. Acta*. 2009, 654, 133–140.
[61] Simone, P. S.; Anderson, G. T.; Emmert, G. L. *Anal. Chim. Acta*. 2006, 570, 259–266.
[62] Bruzzoniti, M. C.; De Carlo, R. M.; Horvath, K.; Perrachon, D.; Prelle, A.; Tófalvi, R.; Sarzanini, C.; Hajós, P. *J. Chromatogr. A*. 2008, 1187, 188–196.
[63] Yoshikawa, K.; Soda, Y.; Sakuragawa, A. *Anal. Sci.* 2009, 25, 1491–1494.
[64] Product Manual IonPac AS24, Document nº 065163, Revision 1.0 (2007). http://www.dionex.com (accessed Dec. 2013).
[65] Ding, Y.; Rogers, K. *Electrophoresis*. 2010, 31, 2602–2607.

[66] Zhang, H.; Zhu, J.; Aranda-Rodriguez, R.; Feng, Y. L. *Anal. Chim. Acta.* 2011, 706, 176–183.

[67] Kubáň, P.; Makarõtševa, N.; Kiplagat, I. K.; Kaljurand, M. *J. Sep. Sci.* 2012, 35, 666–673.

[68] Jang, A.; Zou, Z. W.; Lee, K. K.; Ahn, C. H.; Bishop, P. L. *Meas. Sci. Technol.* 2011, 22, 032001.

[69] Suedee, R.; Intakong, W.; Lieberzeit, P. A.; Wanichapichart, P.; Chooto, P.; Dickert, F. L. *J. Appl. Polym. Sci.* 2007, 106, 3861–3871.

[70] Suedee, R.; Intakong, W.; Dickert, F. L. *Anal. Chim. Acta.* 2006, 569, 66–75.

[71] Suedee, R.; Srichana, T.; Sangpagai, C.; Tunthana, C.; Vanichapichat, P. *Anal. Chim. Acta.* 2004, 504, 89–100.

[72] Suedee, R.; Intakong, W.; Dickert, F. L. *Talanta.* 2006, 70, 194–201.

[73] Geme, G.; Brown, M. A.; Simone, P. S.; Emmert, G. L. *Water Res.* 2005, 39, 3827–3836.

[74] Emmert, G. L.; Geme, G.; Brown, M. A.; Simone, P. S. *Anal. Chim. Acta.* 2009, 656, 1–7.

[75] Zhai, H.; Zhang, X. *Water Res.* 2009, 43, 2093–2100.

[76] Zhang, X.; Talley, J. W.; Boggess, B.; Ding, G. Y.; Birdsell, D. *Environ. Sci. Technol.* 2008, 42, 6598–6603.

In: Drinking Water and Water Management
Editor: Mason Giannino ISBN: 978-1-63117-523-7

Chapter 4

REMOVAL OF HIGH LEVELS OF NATURALLY OCCURRING RADIOACTIVITY FROM DRINKING WATER

A. Salas, A. Baeza, J. Guillén, A. Muñoz and J. G. Muñoz
LARUEX, Dpt. Applied Physics, Faculty of Veterinary Sciences, University of Extremadura, Cáceres, Spain

ABSTRACT

Water used for human consumption can present different types of contaminants, mainly chemical and biological. The routine procedures of Drinking Water Treatment Plants (DWTP) remove the majority of these contaminants. However, the radioactivity content present in the water remains mostly unaffected by these routine procedures. The sources of radionuclides can be either anthropogenic or naturally occurring, the latter being closely related to the geological characteristics. The naturally occurring radionuclide content of surface and ground waters are the result of the incorporation into the original rainwater of diverse chemical elements found in the soils through which the water flows. In the case of areas not affected by serious radioactive contamination events, the natural source is the most important. Therefore, the drinking water intake can pose a radiological hazard, and it is regulated by establishing a series of quantitative thresholds regarding the Total Indicative Dose (TID), and other parameters such as gross alpha and beta content. There are a lot of radiochemical procedures that are able to eliminate practically the main radioactive ions solved in water. Their generalized application is, however, limited by their high selectivity for each element or group of elements and their relatively high costs. It occurs often that the water treated acquires chemical characteristics which make it not drinkable. The traditional procedures of DWTPs can be also modified to maximize the removal of radionuclides, mainly uranium and radium, while the output water remains within drinkable limit, also established by legislation. These modifications are obtained by analyzing the best conditions for all steps in the process, from filtration to coagulation-floculation, for different types of reagents at varying concentrations and pH. Then, they were tested successfully in operating DWTPs, which provided the population with drinkable water which radionuclide content was much lower.

1. INTRODUCTION

Many organic and inorganic contaminants may be found in drinking water. However, an important group of them – the radioactive contaminants such as uranium and radium – are seldom taken into account, even though they can cause health problems even at low concentrations (WHO, 1979). Since drinking water can contain naturally occurring radionuclides in solution, the potential problem needs to be studied from different perspectives since various chemical, physical, and industrial processes may be required before the water is radiologically apt for human consumption. Uranium and radium are usually the most important naturally occurring radionuclides present in water. Thus, Kitto et al. (2005) found significant values of uranium and radium in drinking waters in USA, with radium activities of up to 2 $Bq \cdot L^{-1}$. In Europe, high concentrations of uranium and radium have often been found in waters of Norway, Finland and Germany (Frengstad et. al. 2000, WHO, 2004), and relatively high concentrations (0.01–10) $\mu g \cdot L^{-1}$ of uranium have been found in wells of Northern Greece, (Katsoyiannis, 2007). In Spain, the uranium content was in the range (0.28–18610) $mBq \cdot L^{-1}$ for $^{234,238}U$ and (1.7–718) $mBq \cdot L^{-1}$ for ^{226}Ra. (Herranz et al., 1999; Baeza et al., 2008). The radium content was within the range (1.7-718) $mBq \cdot L^{-1}$ for ^{226}Ra and (1.3-267) $mBq \cdot L^{-1}$ for ^{228}Ra (Manjón et al., 1997; Dueñas et al., 1999; Baeza et al., 1995).

The characteristics of natural surface and ground waters are the result of the incorporation into the original rainwater of diverse chemical elements found in the soils through which the water flows. The radionuclides of the natural radioactive series can be present at relatively high concentrations in granitic and metamorphic rocks such as gneisses, black clays, and phosphate rocks (Suarez et al., 2000). Laboratory experiments to analyze the water-rock interactions in granite showed an increase of the uranium content in water in close contact with granite (Bonotto et al., 2001). The correlation between soils with high levels of natural radioactivity and the activities incorporated into the waters has also been confirmed in different geological environments (Otton, 1994; Baeza et al., 1995; Banks et al., 1998). The connection is of course especially strong for groundwaters (Hess et al., 1985; Orloff et al., 2004).

Agricultural activities constitute another major source of surface water contamination. Over 140 million tons of fertilizers and several million tons of pesticides are applied to soils each year (Schriks et al. 2010). The naturally occurring radionuclides associated with fertilizers can be leached from soil by run-off water into surface and groundwater resources, being this one of the causes of eutrophication of water bodies (Badruzzaman et al., 2012). Other chemical species, such as phosphates, nitrates and sulphates, also showed increased levels in water bodies contaminated by use of fertilizers (Barisic et al., 1992; Zielinski et al., 1997; Baeza et al., 2011). Uranium occurs usually in water bodies contaminated by fertilizers, and the ratio $^{234}U/^{238}U$ has been considered as a potential marker for tracking nutrient sources (Zielinski et al., 1997; Badruzzaman et al., 2012). Increases in the uranium content by factors 1.4-5.3 have been observed in a river receiving the run-off water of farming lands using fertilizers (Baeza et al., 2011).

The content of naturally occurring radionuclides may become an increasingly serious problem for drinking water supplies, due to the growing demand for new water resources (IAEA, 2003a). Unless the corresponding water treatment systems are appropriately designed, a significant part of this radioactive content can eventually be incorporated into the population, with the consequent risk to health.

2. Determination of Radionuclides

Radionuclides are unstable isotopes that usually lose their excess energy by the emission of ionizing radiation. The emission probability is different for each radionuclide and is expressed in terms of its half-life. The half-life of a radionuclide, $T_{½}$, is the time it takes for the activity to decrease by a half. There are three main types of ionizing radiation: α, β, and γ, whose main properties are the following:

1. α-particles are helium nuclei (2 protons + 2 neutrons). They have high values of mass and kinetic energy, the latter within the range 3.95–8.78 MeV[1]. Their interaction with matter is so strong that even a sheet of paper is enough to stop them, with all their energy being lost in crossing its thickness. Most of α-emitting radionuclides are naturally occurring: ^{238}U, ^{226}Ra, ^{210}Po, ^{232}Th, etc.
2. β-particles have various origins depending on the characteristics of the nucleus (β^-, β^+, EC or electron conversion, …), but the final result is the emission of electrons or positrons within a wide range of energies (0.01859–3.54 MeV). Due to their low mass and charge compared with α-particles, they can penetrate more deeply into matter, requiring a few centimetres of aluminium to stop them. Some β-radionuclides are: ^{3}H, ^{210}Pb, ^{90}Sr, ^{40}K, ^{137}Cs etc.
3. γ-radiation consists of high-energy photons, from a few keV to less than 10 MeV. Their penetrating power is greater than the previous particles, and they require several centimetres of lead to be stopped. This type radiation is frequently emitted immediately after α- or β-radiation. ^{40}K, ^{137}Cs, and ^{226}Ra, among others, are also γ-emitting radionuclides.

Systems to detect radionuclides are different for each type of radiation, because they have different characteristics. Due to the low penetrating power of α and β particles, it is necessary to perform radiochemical separation procedures for their proper assay. In the case of γ-radiation, however, such radiochemical separations are usually unnecessary. The determination of α-emitting radionuclides is usually by means of silicon detectors within a vacuum chamber. Gas-flow proportional counters and LSC (liquid scintillation counting) are used to determine both β-particles and α-particles. The determination of γ-emitting radionuclides is by means of NaI(Tl) and HP(Ge) crystals.

[1] An electron volt (eV) is a unit of energy equal to the amount of kinetic energy gained by a single unbound electron when it accelerates through an electric potential difference of 1 V. 1 eV = $1.609 \cdot 10^{-9}$ J.

3. NATURALLY OCCURRING RADIONUCLIDES IN THE ENVIRONMENT

The naturally occurring radionuclides comprise those belonging to the natural radioactive decay series, ^{40}K, and cosmogenic radionuclides, which are formed by the interaction of high-energy cosmic rays with atomic nuclei in the atmosphere. There are three natural radioactive series present in nature: thorium, uranium, and actinium, also denominated A = 4n, A = 4n+2, and A = 4n+3, where A is the mass number of the radionuclide. These natural series begin with ^{232}Th (thorium series), ^{238}U (uranium series), and ^{235}U (actinium series), and finish with ^{208}Pb, ^{206}Pb, and ^{207}Pb, respectively. The half-lives of these radionuclides are usually very large, $4.468 \cdot 10^9$, $7.04 \cdot 10^8$, and $1.405 \cdot 10^{10}$ y for ^{238}U, ^{235}U, and ^{232}Th respectively Figure 1 shows the decay chain of the uranium series as a way of example.

Uranium, as it occurs in nature, consists of a mixture of the three main radioisotopes ^{238}U, ^{235}U, and ^{234}U. The natural abundance of the uranium isotopes is for ^{234}U (0.005±0.001)%, (0.720±0.001)% for ^{235}U and (99.275±0.002)% for ^{238}U. Radium has three naturally occurring isotopes that are usually present in drinking water. These are ^{226}Ra, with half live of 1600 y, ^{228}Ra, with half live of 5.7 y and ^{224}Ra, with half live of 3.64 d. ^{226}Ra is an α-emitter belonging to the uranium series (see figure 1), and it is usually the most abundant radium isotope in water. ^{228}Ra is a β-emitter in the ^{232} Th decay chain. Its second generation daughter is ^{224}Ra, an alpha emitter that is expected to be present in water that contains its progenitor, ^{228}Ra. Thorium is a naturally occurring radioactive member of the actinide series of thorium and uranium. It is composed of 25 different isotopes but the most important in water are ^{230}Th with half live of $8 \cdot 10^4$ y, ^{232}Th, half live of $1.41 \cdot 10^{10}$ y and ^{228}Th, half live of 1.91 y. Polonium is one of the rarest naturally occurring elements ($3 \cdot 10^{-14}$% of the lithosphere). There are a great number of polonium radioisotopes produced in the natural decay series. The most important of them, regarding its presence in water, is ^{210}Po ($T_{½}$ = 138.4 d), which is a descendant of ^{210}Pb and belonging to the uranium series. Finally, the isotope ^{40}K ($T_{½} = 1.277 \cdot 10^9$ y) is also a naturally occurring radionuclide with a long half-life, and constitutes 0.012% of natural abundance of potassium.

The distribution of these radionuclides in the geosphere depends on the distribution of the geological media from which they derive and the processes which concentrate them at a specific location in specific media (IAEA, 2003b).

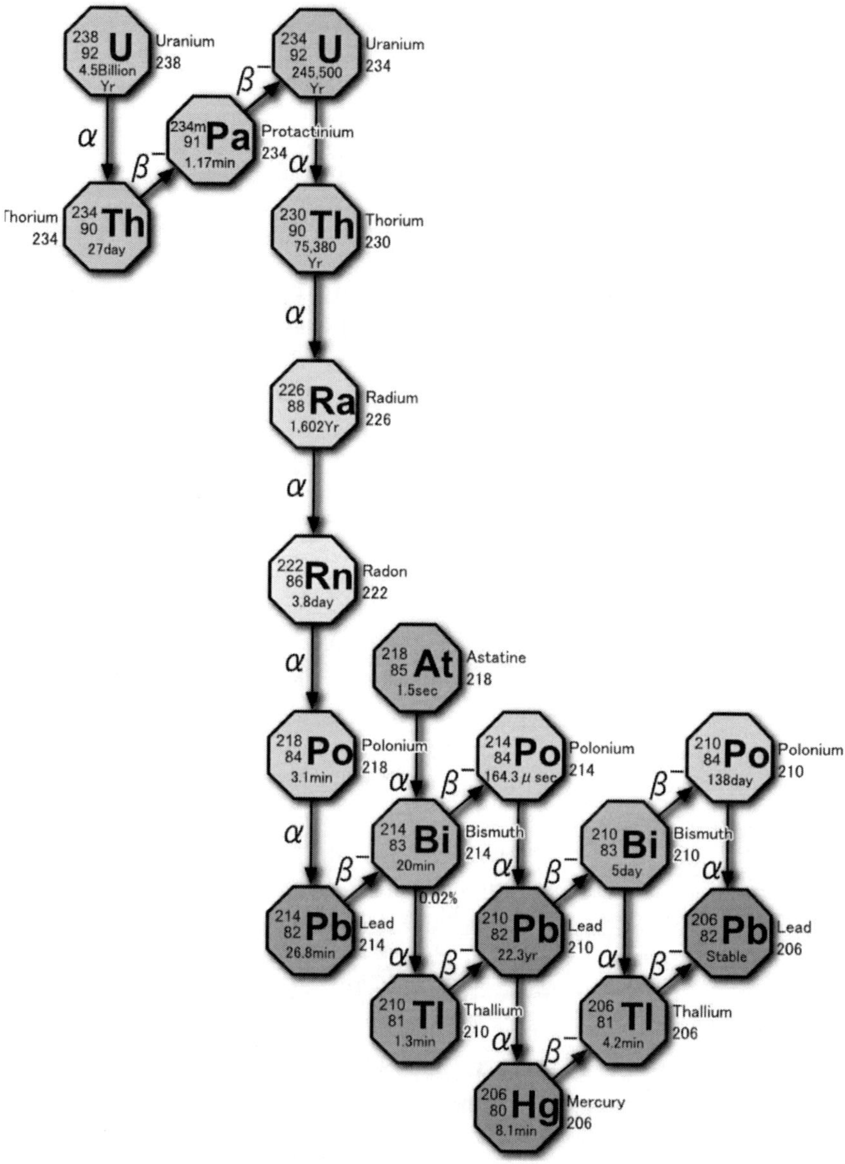

Graphic from (Tosaka, 2008).

Figure 1. Decay chain of the natural uranium series (4n+2).

They can also be enhanced locally by human activities, and may be a cause for significant radiological concern because of NORM (Naturally Occurring Radioactive Material) industries. These industries use raw materials with high contents of natural radionuclides, and their bioavailability in their products, by-products, and residues may be enhanced due to physicochemical changes or to the form in which they are managed. Mining industries (iron, copper, …), the production of fertilizers from phosphate rocks, and burning non-nuclear fuels (coal, oil, gas) are some examples of NORM activities (IAEA, 2003b).

4. Chemical Speciation of Naturally Occurring Radionuclides in Water

Naturally occurring radionuclides content in water are at trace levels. The inorganic chemical compounds, which are also present in the water as ions but at higher concentrations, condition the occurrence of different chemical species for a given radionuclide. These species can have physico-chemical properties that can affect the adsorption or co-precipitation of radionuclides from water, and therefore their removal. The determination of the chemical speciation of inorganic compounds of naturally occurring radionuclides solved in water was carried out by using the chemical constants included in geochemical indices (Wolery, 1992).

Uranium have four oxidation states usually represented by U^{3+}, U^{4+}, UO_2^{+} and UO_2^{2+} (+6). The uranyl ion (UO_2^{2+}) is the most common species in water, and in some cases the only one. It can be easily combined with Cl^-, NO_3^-, SO_4^{2-} and CO_3^{2-} ions. Figure 2 shows the distribution of uranyl inorganic compounds depending on pH in two types of water from different origin: groundwater and water surface. The predominant uranyl species depended on pH (see Figure 2). At pH close to 6, the neutral species of carbonates or hydroxides are the most significant for drinking water- UO_2CO_3(aq) and $UO_2(OH)_2$(aq). Between pH 7-10, the soluble carbonate complexes of uranyl are the predominant anion species- $UO_2(CO_3)_2^{2-}$ and $UO_2(CO_3)_2^{4-}$. Above pH 11 and in carbonate-free water, hydroxide complexes (anion) are formed, $UO_2(OH)_3^-$ and $UO_2(OH)_4^{2-}$. The content of inorganic complexes, which was different for each kind of water, also influenced significantly the speciation of uranyl compounds. So, for low carbonate concentration, as occurs in surface water, the species of hydroxide and sulphate uranyl become more relevant. At pH < 5, the predominant species in the groundwater was UO^{2+}, while for surface water was UO_2SO_4(aq). At pHs 5-6, the predominant species in groundwater was UO_2CO_3(aq) while for surface water the predominant species were UO_2CO_3(aq) and $UO_2(OH)_2$(aq). At pH > 7, there were not many differences for the two types of water.

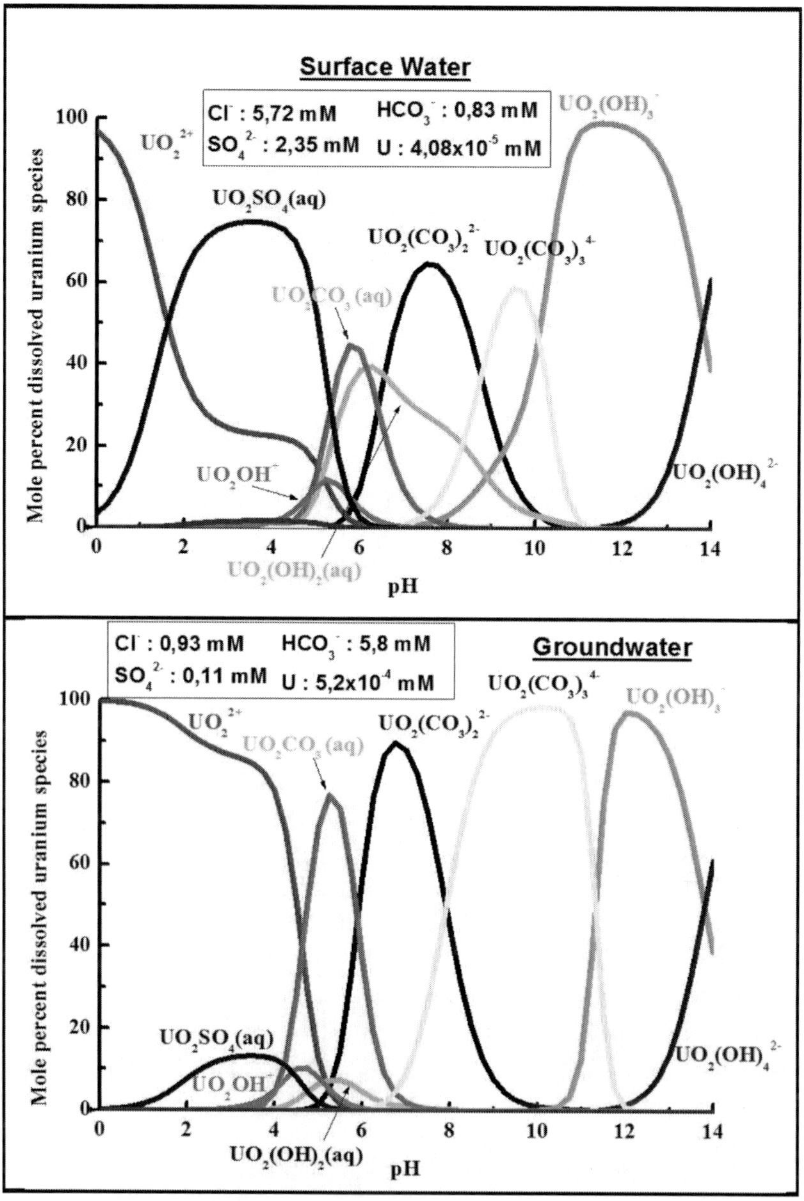

Figure 2. Distribution of uranyl species complexes solved in surface water (up) and groundwater (down) versus pH.

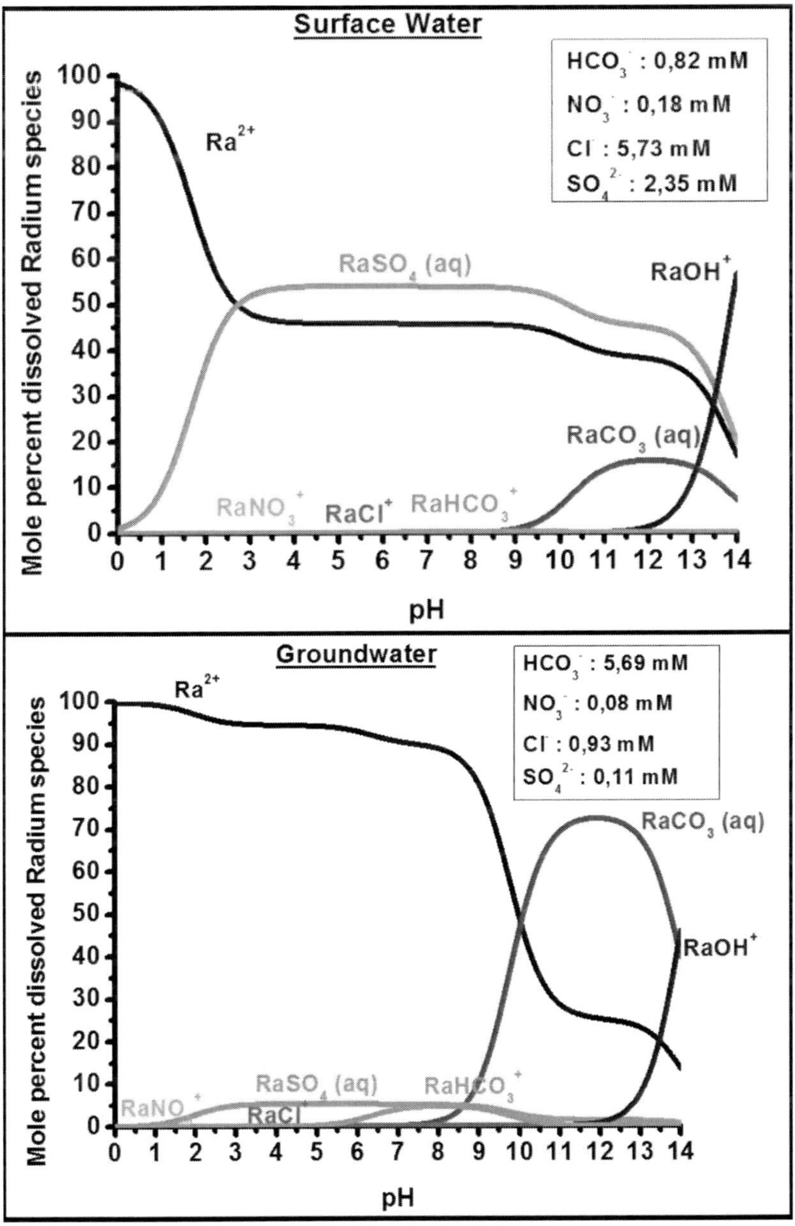

Figure 3. Distribution of radium species complexes solved in surface water (up) and groundwater (down) versus pH.

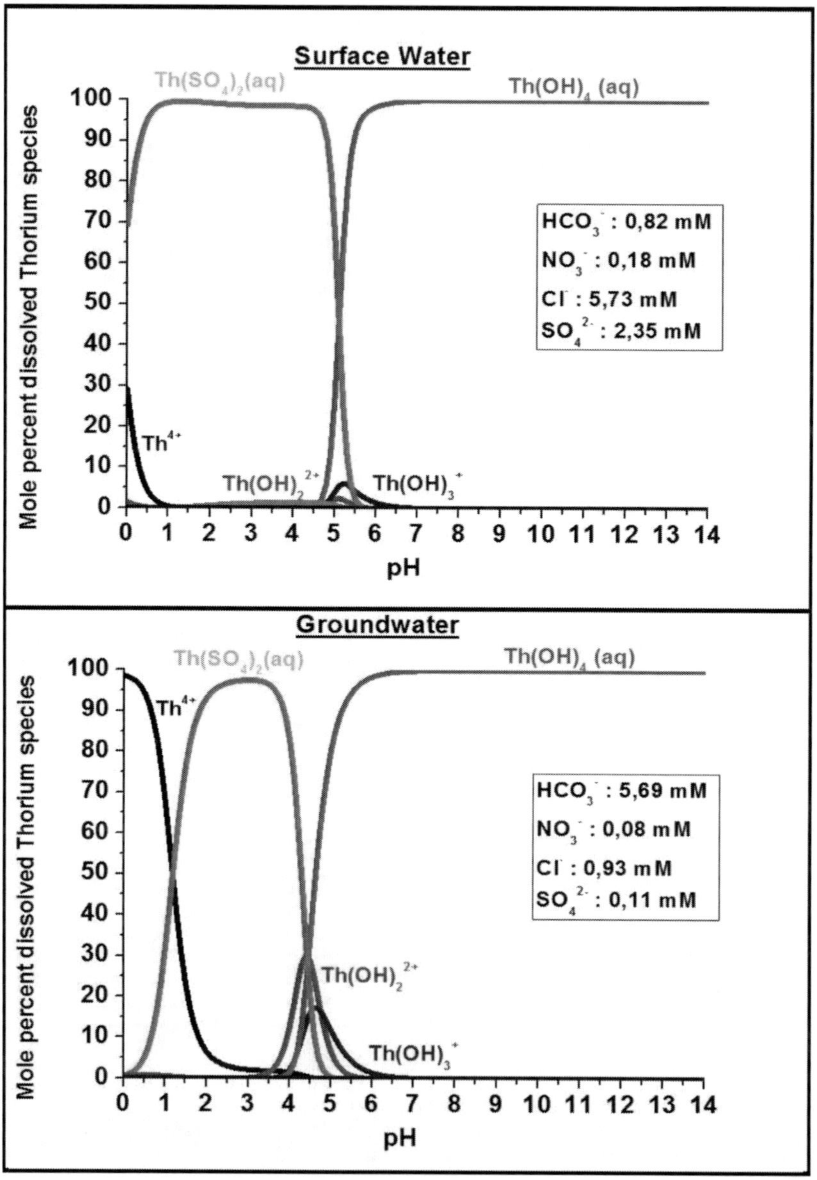

Figure 4. Distribution of thorium species complexes solved in surface water (up) and groundwater (down) versus pH.

Radium is the largest of the group IIA alkaline earth metals, and its chemistry is similar that of barium. This similarity has been used to co-precipate it as Ba(Ra)SO_4 in very acidic solutions (Baeza 1998). Figure 3 shows the distribution of radium species versus pH in surface water and groundwater. The main difference between the two types of water was their sulphate and bicarbonate contents. When the predominant anions were sulphates, surface water, for all pHs, the radium species complexes are Ra^{2+} and $RaSO_4(aq)$, which is very soluble and normally does not precipitate by itself. In ground water the Ra^{2+} is the most abundant at pH < 10, and it is $RaCO_3(aq)$ for higher values.

Thorium has low mobility, mainly due to the high stability of the insoluble oxide ThO_2, with the result that the concentration of soluble thorium in water will be very low (Atwood, 2010). Figure 4 shows the distribution of the low content of thorium soluble in water. Two species are dominant in water, $Th(SO_4)_2(aq)$ at pH < 5, and $Th(OH)_4(aq)$ at pH > 5, regardless of the water origin, surface water or groundwater.

Polonium was predicted to have stable oxidation states of -2, +2, +4, and +6, based on analogy with Se and Te, (Figgins, 1961). The evidence for polonium (+6) is still rather slight, the other states are well established with the +4 oxidation state being the stable state in solution. Due to the lack of data of formation constants of polonium compounds, no modelization of the speciation in water was possible. However, there is some information about environmental behaviour of polonium in water, especially seawater. In oxygenated seawater, ^{210}Po is most likely to exist as oxides, which are highly insoluble (Sabuti et al., 2010), although the formation of carbonate complexes may render ^{210}Pb more soluble (Platford et al., 1986). The ^{210}Po content is closely related to ^{210}Pb. It also shows a high affinity for particulate matter, greater than that observed for ^{210}Pb (Carvalho, 1995).

5. Removal of Naturally Occurring Radionuclides from Drinking Water

Waters usually have to undergo a series of processes known as potabilization to satisfy physico-chemical and biological criteria for them to be considered apt for human consumption (EU 1998; BOE, 2003). European legislation provides parametric values for alpha and beta emitters, and the Total Indicated Dose (TID), (see Table 1). The gross alpha and beta indices can be considered as an estimate of alpha emitters and beta present in the water, while the TID estimate the annual effective dose due to the consumption of a given water. TID calculation is performed according to equation (1),

$$\text{TID}\ (\text{Sv}\cdot\text{y}^{-1}) = V_{j\text{-}a} \sum I_i\, F(g)_{i,\, j\text{-}a} \qquad \text{(eq. 1)}$$

where $V_{j\text{-}a}$ is the average annual volume of water ($L\cdot y^{-1}$) ingested by a young or adult individual, I_i is the activity ($Bq\cdot L^{-1}$) of radionuclide i in solution, $F(g)_{i,j\text{-}a}$ is the conversion coefficient ($Sv\cdot Bq^{-1}$) for each radionuclide i, depending on whether the individual considered young or adult (ICRP, 2012).

There are many different types of drinking water treatment plants (DWTP), and their complexity depends on the size of the population to be supplied and the type of raw water.

Table 1. Parametric values for alpha and beta emitters, and the Total Indicated Dose (TID) (EU, 1998)

Parameter	Parametric value	Notes
Total Indicative Dose (TID)	0.10 $mSv\cdot y^{-1}$	Excluding tritium, ^{40}K, radon, and radon decay products
Gross alpha Activity	0.1 $Bq\cdot L^{-1}$	
Gross beta Activity	1 $Bq\cdot L^{-1}$	Excluding tritium and ^{40}K
Tritium	100 $Bq\cdot L^{-1}$	

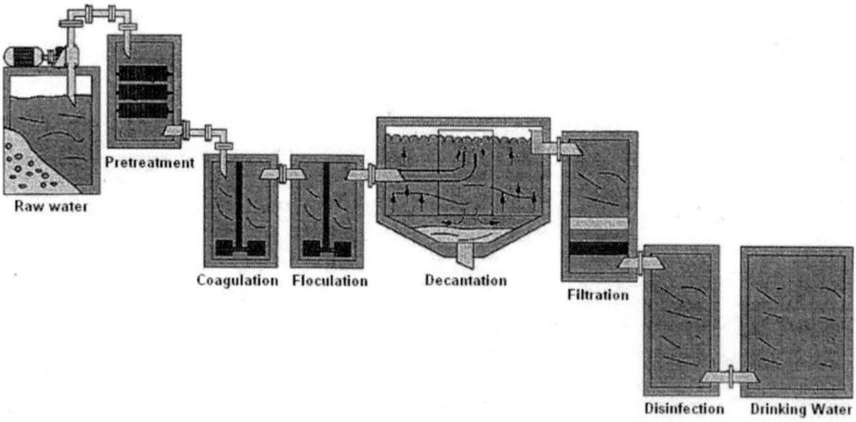

Figure 5. Schematic diagram of the different processed carried out in a DWTP.

The DWTPs use a series of chemical reagents adapted to the particular characteristics of the raw water being treated. Figure 5 shows a schematic diagram of the processes carried out in a typical DWTP. After an initial pre-treatment step, in the coagulation–floculation step coagulants are used to neutralize colloidal particles in the water and precipitate them in decantation tanks, forming thus sludge as a by-product. Then the water is filtrated, usually through sand beds. Finally, it is disinfected in order to remove biological hazards, and supplied to the population. In certain types of water treatment plants, there are adsorbents that can improve water quality. These processes are supervised in order to meet the adequate parametric values according to the corresponding legislation.

5.1. Specific Methods

There are many methods for elimination of radionuclides from water by specific materials. These methods obtained good results for the removal of the naturally occurring radionuclides studied, but their working condition may not so compatible with parameters and criteria uses in DWTP, such as pH of end water, volumes of reagents used, costs, etc. High elimination rates can be obtained with the use of very selective substances such as ion exchange resins. In particular, some cation exchange resins give the best results for acidic pHs (pH 3.5), and some anion exchange resins also give high uranium elimination percentages, close to 100%, depending on the specific bed-volumes of resin used (Lee et al., 1982; Lee et al., 1983; Hanson et al., 1987; Hathaway, 1983; Utrecht, 1985). For radium elimination percentages of the order of 80–98% can be achieved by cation exchange (Brink et al., 1978). There are also diverse types of resins and adsorbent materials have been developed, impregnated with substances that have a great radium adsorbing power, such as MnO_2-coated acrylic fibre filters (Moorc, 1978), the Radium Selective Complexer, RSC (Dow, 1986) and $BaSO_4$-impregnated alumina (Clifford et al., 1988).

Another type of purification procedures that are able to effectively remove radionuclides from water are those employing reverse osmosis treatment. This type of treatment is often used when there is lack of water resources, but there is seawater or water with high concentration of salts. Diverse studies for the removal of a number of inorganic contaminants and uranium by reverse osmosis showed that the natural uranium and radium was highly removed (99%) by different reverse osmosis membranes. This method was very effective for uranium and radium removal (Huxstep et al., 1987; Sorg et al.,

1980). Recently, this treatment results has been achieved very high removal, 90% for alpha emitters, beta and uranium in surface water (Montaña et al., 2013), or 99% in groundwater (Khert, 2013).

There have also been several studies of the elimination of uranium and radium from water destined for public consumption with a lime-softening treatment. At pHs greater than 10, the use of lime plus magnesium carbonate has been observed to eliminate nearly 100% of uranium (Logsdon, 1997). This was due to the fact that basic pHs causes destabilization of stable cations which help the co-precipitation of radionuclides present in the water. Also, the use of lime at pHs above 10 provokes the precipitation of other cations in the form of salts, contributing in parallel to the elimination of radium.

The use of activated carbon is considered to be a successfully available technology for removing low-solubility contaminants in water treatment, including trace metals (Derbyshire et al., 2000; Bansal et al., 2005). Different studies of radionuclides removal using carbon adsorbents have been carried out. Activated carbon can be applied to the removal of uranium from aqueous solutions with adsorption efficiency 92%, at pH 5 (Abbasi 1994; Kütahyal 2004). In the case of radium, Miró et al. 2008 studied the adsorption of ^{226}Ra from natural water by a granulated wood charcoal as a function of the solution pH, in the range 4–10. In this range, the adsorption efficiency of ^{226}Ra increased from 40% up to 80%, fitting a parabolic function. These results were explained by considering the different species of each radionuclide present at the pH values of solution at the end of the adsorbent column, and the pH of the point of zero charge of the wood charcoal. However, the use of activated charcoal has the disavandtage that it is an unspecific adsorbent. Therefore, in the case of waters with significant concentrations of inorganic cations and anions in solution, the adsorption of trace elements in them, such as uranium and radium, can be adversely affected (Salas 2005).

5.2. Conventional and Modified Potabilization Treatment

Radioactive decontamination of water during the water treatment process has been an issue repeatedly studied by various authors, whose goal was to analyze how the treatment process influences the radioactive decontamination of waters (Lee 1982; Sorg 1990; Real 2000). However, the results achieved depended on the working conditions of the DWTP considered. Regarding the traditional potabilization processes, the coagulation-floculation step seemed to be the best suited process to remove radionuclides from the water. The most commonly coagulants used are based on iron or aluminium (Degremont, 1979). Indeed, with respect to uranium decontamination, Lee et al. (1983), found that, with only low doses (25 $mg \cdot L^{-1}$ or less) of such reagents as ferric sulfate, ferrous sulfate, or aluminium sulfate, uranium was removed from drinking water by percentages greater than 80% for water pHs 6 and 10. The author suggested that the pH dependency of uranium removal was related to the stability and charge characteristics of the uranyl species and the metal hydroxide precipitates at the adjusted pH of the solution. For example, at pH 6 the predominant complex would be the neutral $UO_2CO_3(aq)$ (figure 2) and the charge of the hydroxide precipitates would also be near zero. Baeza et al. (2008) tested the concentration ranges, types of coagulants-floculants and pHs most commonly used in DWTPs for the removal in natural water of uranium and radium. They found that the decontamination obtained for both radionuclides to be independent of the coagulant dose used. The pH at which the process of coagulation was carried out, however, was found to be a determining factor in the effectiveness of the elimination of both uranium and radium. In the case of uranium, the best elimination percentages, in all cases greater than or equal to 80%, were found for coagulation at pH 6, independently of which particular coagulant was used. In the case of radium, the elimination was greater as the pH was more basic, reaching at pH 10 efficiencies of the order of 25% for the aluminium-based reagents and of 50% for iron-based reagents, the iron-based coagulants yielding the best values of elimination. In this sense, Gäfvert et al. (2002) studied a waterworks that used surface water. The authors observed a high removal capacity for uranium (about 85%), thorium (90%), plutonium (95%) and polonium (90%) in the coagulation–flocculation process at pH about 6 with $Al_2(SO_4)_3$ and $FeCl_3$, while caesium, strontium and radium pass through the purification process with almost unchanged activity concentrations.

The pH in the coagulation process in a conventional DWTP is generally similar to the raw water, usually in the range 7-8.5, because it will eventually be the final pH of the water to be supplied to the population. Baeza et al. (2012) studied the modification of the routine working conditions of an operating drinking water treatment plant in order to maximize uranium removal. Under routine conditions, the uranium and radium removal was practically negligible. This routine procedure can be modified to remove naturally occurring radionuclides. Figure 6 shows the modified uranium removal process for surface water in an operating DWTP. It can be seen that adjustments of the pH by adding HCl were made in the mixing zone to reach pH 6, in order to optimize the uranium removal.

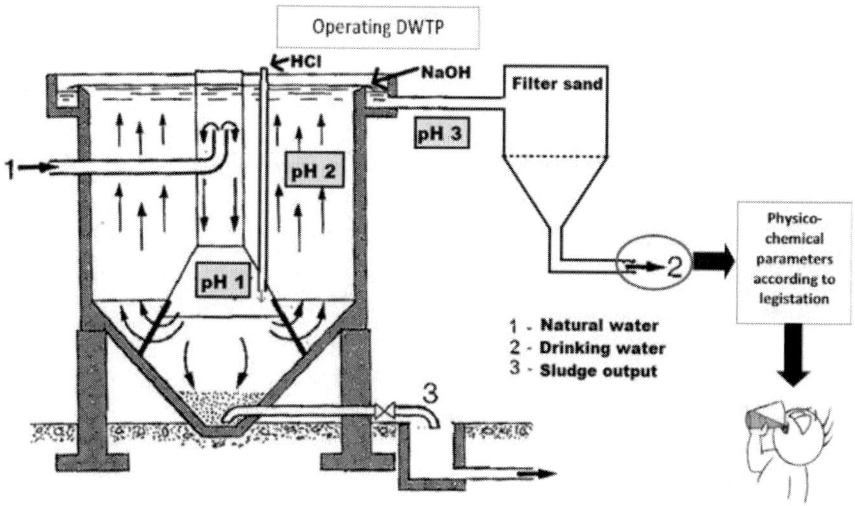

Figure 6. Schematic diagram of the modification of an operating DWTP to maximize the uranium removal.

Then, NaOH was added in the clarifier zone so as to reach the final pH 7. Finally the water is stored until distribution to the population. Using the modified procedure, the gross alpha activity was reduced below the parametric value of 0.1 $Bq{\cdot}L^{-1}$ and the uranium removal efficiencies were in the range 53–83 %. The use of the modified procedure did not change significantly the physicochemical parameters of the drinking water regarding its corresponding natural water. Subsequently, the authors applied this method to other DWTPs using groundwater: two with high concentrations of uranium and radium another. In the latter, the reagents were used in reverse order, first NaOH and then HCl. The results obtained for the TID in the three DWTP are lower than the parametric value, 0.10 $mSv{\cdot}y^{-1}$ (Baeza et al., 2013).

CONCLUSION

Water can contain naturally occurring contaminants that are necessary to remove before it is consumed by living beings. Among these undesirable substances, radionuclides are present at trace level but may have a very high toxicity. Uranium and radium are usually the radionuclides that present higher activity levels in raw water. The radioactivity content in water for public consumption was recently considered in the legislation of some countries, establishing parametric values for gross alpha and beta indices, tritium, and TID. In some locations, raw water can exceed these parameters. Therefore, the radionuclide removal from water during the potabilization process is required in order to produce drinkable water. Conventional methods of potabilization, using pH above 6 in the coagulation process or processes that employ only disinfection are ineffective for the removal of these radionuclides. The content of inorganic compounds and the pH of the raw water are a key elements to understand the behaviour of naturally occurring radionuclides in water, and their possible removal. These variables affect the speciation of inorganic compounds of radionuclides that are present in raw water, and therefore, their physico-chemical properties. Two different approaches have been used for radionuclide removal: the use of specific materials or adsorbant and the modification of the working procedure of DWTP. The use adsorbants, such as resins, activated charcoal, and specific materials in the case of radium, presented high elimination rates. However, they have the disadvantage that their use may be costly, and their use by themselves do not guarantee drinkable water. The DWTP based on reverse osmosis procedures have good elimination rates, although the energy cost of the process is high. They are a good option in the case of raw water with high salinity or from marine origin. Finally, the modification of the working conditions can be a good for raw water from surface or groundwater origin. It is based on the modification of the pH condition in which the coagulation-floculation step is carried out. The pH was adjusted to 6 for uranium removal and about 10 for radium removal. They have the disadvantage that additional consumption of acids and bases are required because of pH regulation at the coagulation-floculation and final steps. On the other hand, they have been proved successfully in operating DWTP, and the resulting water was apt for human consumption with reduced levels of naturally occurring radionuclides. Finally, it has to be taken into consideration that the radionuclides successfully removed from drinking water do not disappear, but they are accumulated into by-products, such as sludges and brines, which are to be treated adequately.

ACKNOWLEDGMENTS

This work was made possible by the partial financing provided by the Spanish Nuclear Safety Board (CSN) to the Project "Delimitar el impacto de la potabilización radiológica del agua y probar la capacidad de materiales para la eliminación de radio de las aguas".

References

Abbasi, W., Streat, M., 1994. Adsorption of uranium from aqueous solutions using activitated carbon. *Sep. Sci. Technol.* 29, 1217–1230.

Atwood David A., 2010. Radionuclides in the Environment. Ed. John Wiley & Sons Ltd., 247-252.

Badruzzaman M., Pinzon J., Oppenhaimer J., Jacabgeli J. G., 2012. Sources of nutrients impacting surface waters in Florida: a review. *J. Environ. Manag.* 109, 80-92.

Baeza, A., del Río, L. M., Jiménez, A., Miró, C., Paniagua, J. M., 1995. Factors determining the radioactivity levels of waters in the province of Cáceres (Spain). *Appl. Radiat. Isot.* 46(10), 1053-1059.

Baeza A, del Río LM, Jiménez A., 1998. Procedure for simultaneous determination of $^{223,\ 224,\ 226,\ 228}$Ra by Alpha and Gamma Spectrometry. *Radiochim. Acta* 83, 53–60.

Baeza, A. Salas, F. Legarda, 2008. Determining factors in the elimination of uranium and radium from groundwaters during a standard potabilization process, *Sci. Total Environ.* 406 24-34.

Baeza A., Corbacho J. A., Guillen J., Salas A., Mora J. C., 2011. Analysis of the different source terms of natural radionuclides in a river affected by NORM (Naturally Occurring Radioactive Materials) activities. *Chemosphere* 83, 933–940.

Baeza A., Salas A., Guillén J., 2012. Adaptation of working conditions of an operating Drinking Water Treatment Plant to remove of naturally occurring radionuclides. *Water Air Soil Poll.* 223, 5057-5069.

Baeza A., Salas A., Guillén J., Muñoz-Serrano A., 2013. Association of naturally occurring radionuclides in sludges from Drinking Water Treatment Plants previously optimized for their removal. *Chemosphere*, http://dx.doi.org/10.1016/j.chemosphere.2013.10.040 (in press)

Banks, D., Reimann, C., Skarphagen, H., 1998. The comparative hydrochemistry of two granitic island aquifers: the Isles of Stilly, UK and the Hvaler Islands, Norway. *Sci. Total Environ.* 209, 169–83.

Bansal, R. C., Goyal, M., 2005. Activated Carbon Adsorption. Taylor & Francis, Boca Raton.

Barisic, D., Lucic, S., Miletic, P., 1992. Radium and uranium in phosphate fertilizers and their impact on the radioactivity of waters. *Water Res.* 20(5), 607–611.

BOE, Boletín Oficial del Estado. Criterios sanitarios de la calidad del agua de consumo humano. Real Decreto 140/2003 de 7 de febrero, Madrid, Spain, vol. 45. 2003. p. 7228–45. (in Spanish)

Bonotto D. M., Andrews J. N., Barbyshire D. P. F., 2001. A laboratory study of the transfer of ^{234}U and ^{238}U during water-rock interactions in the Carnmenellis granite (Cornwall, England) and implications for the interpretation of field data. *Appl. Radiat. Isotop.* 54, 977-994.

Brink, W. L., Schliekelman, R. J., Bennett, D. L., Bell, C. R., Markwood, I. M., 1978. Radium removal efficiencies in water treatment processes. *J. Am. Water Works Assoc.* 70(1), 31.

Carvalho F. P., 1995. ^{210}Pb and ^{210}Po in sediments and suspended matter in the Tagus estuary, Portugal. Local enhancement of natural levels by wastes from phosphate ore processing industry. *Sci. Total Environ.* 159, 201-214.

Clifford, D., Vijjeswarapu,W. and Subramonian, S., 1988. Evaluating various adsorbents and membranes for Removing Radium from Groundwater. *J. Am. Water Works Assoc.* 80(7), 94–104.

Degremont, 1979. Manual técnico del agua, Rueil-Malmaison CEDEX, France, ed. GRAFO S. A., Bilbao, 135–143 pp. (in Spanish)

Derbyshire, F., Jagtoyen, M., Andrews, R., Rao, A., Martín-Guyón, I., Grulke, E. A., 2000. Carbon materials in environmental applications. In: Radovic, L. R. (Ed.), *Chemistry and physics of carbon*, vol. 27. Marcel Dekker, New York, pp. 1–66.

Dow Chemical Company, 1986. Material safety data sheet: XFS 43230.00 Experimental Radium Complexer, Midland, MI.

Dueñas C, Fernández MC, Carretero J, Liger E, Cañete S., 1999. ^{226}Ra and ^{222}Rn concentrations and doses in bottled waters in Spain. *J. Environ. Radioact.* 45, 283–90.

EU, 1998. European Commission Directive, 98/83/EC, related with drinking water quality intended for human consumption, Brusels, Begium, 3/11/1998; L330: 32-54.

Frengstad B, Midtgard Skrede Aa, Banks D, Krog JR, Siewers U., 2000. The chemistry of Norwegian groundwaters III. The distribution of trace elements in 476 crystalline bedrock groundwaters as analyzed by ICP-MS techniques. *Sci. Total Environ.* 246, 21–40.

Figgins, P. E., 1961. The Radiochemistry of Polonium. National Academy of Sciences Nuclear Science Series 3037.

Gäfvert T., Ellmark C., Holm E., 2002. Removal of radionuclides at a waterworks. *J. Environ. Radioactiv.* 63, 105–115.

Hanson, S. W., Wilson, D. B. and Gunaji, N. N., 1987. Removal of uranium from drinking water by ion exchange and chemical clarification. U.S. EPA Report 600/52-87/076.

Hathaway, S. W., 1983. Uranium Removal Processes. Paper presented at the AWWA Annual Conference, Las Vegas, NE.

Herranz M, Abelairas A, Legarda F., 1999. Uranium contents in raw waters from Biscay (Spain). *Appl. Radiat. Isot.* 51, 203–8.

Hess, C. T., Michel, J., Horton, T. R., Prichard, H. M., Coniglio, W. A., 1985. The occurrence of radioactivity in public water supplies in the United States. *Health Phys*. 48, 553–86.

Huxstep, M. R., Sorg, T. F., 1987. Reverse Osmosis Treatment to Remove Inorganic Contaminants from Drinking Water. U.S. EPA Report EPA/600/52-87/109.

IAEA, 2003a. International Atomic Energy Association. Aguas turbulentas: cambiando el curso. Boletín del OIEA 45, 33–40. (in Spanish)

IAEA, 2003b. Extent of Environmental Contamination by Naturally Occurring Radioactive Material (NORM) and Technological Options for Mitigation. IAEA Technical reports series nº 419, Vienna.

ICRP, 2012. Compendium of Dose Coefficients based on ICRP Publication 60. ICRP Publication 119 *Annals of ICRP*, 41(Suppl.).

Khedrt M. G., 2013. Radioactive contamination of groundwater, special aspects and advantages of removal by reverse osmosis and nanofiltration. *Desalination* 321, 47–54.

Kitto ME, Parekh PP, Torres MA, Schneider D., 2005. Radionuclide and Chemical concentrations in mineral waters at Saratoga Springs, New York. *J. Environ. Radioact*. 80, 327–39.

Katsoyiannis IA., 2007. Carbonate effects and pH-dependence of uranium sorption onto bacteriogenic iron oxides: Kinetic and equilibrium studies. *J. Hazard Mater.* B 139, 31–7.

Kütahyal C., Eral M., 2004. Selective adsorption of uranium from aqueous solutions using activated carbon prepared from charcoal by chemical activation, *Sep. Purif. Technol*. 40, 109–114.

Lee SY, Hall SK, Bondietti EA., 1982. Methods of removing uranium from drinking water: ii. present municipal water treatment and potential removal methods. Office of Drinking Water, U.S. EPA, 570/9–82/003.

Lee SY, Bondietti EA, 1983. Removing uranium from drinking water by metal hydroxides and anion exchange resin. *J. Am. Water Works Assoc*. 75(10), 536–40.

Logsdon GS., 1997. Treatment techniques for the removal of radioactive contaminants from drinking water. Manual of Treatment Techniques for Meeting the Interim Primary Drinking Water Regulations, Cincinnati, OH.

Manjón G, Vioque I, Oreno H, García-Tenorio R, García-León M., 1997. Determination of ^{226}Ra and ^{224}Ra in drinking waters by liquid scintillation counting. *Appl. Radiat. Isotopes* 48, 535–40.

Montaña M., A. Camacho, I. Serrano, R. Devesa, L. Matia, I. Vallés, 2013. Removal of radionuclides in drinking water by membrane treatment using ultrafiltration, reverse osmosis and electrodialysis reversal. *J. Environ. Radioactiv.* 125, 86-92.

Miró C., Baeza A., Salas A., Pastor-Valle J. F., Pastor-Villegas J., 2008. Adsorption of ^{241}Am and ^{226}Ra from natural water by wood charcoal. *Appl. Radiat. Isot.* 66, 95–102.

Moore, W.S., 1978. Preparing Manganese Oxide Coated Acrylic Giber and Article There from. United States Patent No. 4.087.853.

Orloff, K. G., Mistry, K., Charp, P., Metcalf, S., Marino, R., Shelly, T., Melaro, E., Donohoe, A. M., Jones, R. L., 2004. Human exposure to uranium in groundwater. *Environ. Res.* 94, 319–26.

Otton, J. K., 1994. Natural Radioactivity in the Environment. Technical report, U.S. Geological Survey, Energy Resource Surveys Program.

Platford R. F., Joshi S. R. J., 1986. The chemistry of uranium and related radionuclides in lake Ontario waters. *Radioanal. Nucl. Chem.* 106(6), 333.

Real, J., Bourrasse, M., Feuerstein, J., Rouxel, R., 2000. Influence des techniques de potabilisation sur la qualité radiologique de l'eau. *Radioprotection*, 35, 31–44. (in French).

Salas A. Study of the removal efficiency of natural radionuclides in compatible with the processes of water potabilization. PhD Thesis, Physics Applied, University of Extremadura, 2005.

Sabuti A. A., Mohaned C. A. R., 2010. Lead. In: Atwood D.A. Radionuclides in the Environment. John Wiley & Sons, UK, 139-147.

Schriks, M., Heringa, M. B., van der Kooi, M. M. E., de Voogt, P., van Wezel, A. P., 2010. Toxicological relevance of emerging contaminants for drinking water quality. *Water Res.* 44, 461–476.

Sorg, T. J., Forbes R. W., Chambers D. S., 1980. Removal of Radium-226 from Sarasota County, Fla., Drinking Water by Reverse osmosis. *J. Am. Water Works Assoc.* 72(4), 230.

Sorg T. J., 1990. Removal of Uranium from drinking water by conventional treatment methods. In: Cothen CR, Rebers PA, editors. Radon, Radium and Uranium in drinking water. Lewis Pu. INC.

Suarez, E., Fernández, J. A., Baeza, A., Moro, Mª. C., García, D., Moreno, J., Lanaja, J. Mª., 2000. Proyecto Marna. Mapa de radiación gamma natural". Colección de Informes Técnicos 5. Consejo de Seguridad Nuclear. (in Spanish).

Tosaka, 2008, Decay chain graphic, Wikipedia, accessed 13.09.2013.

Utrecht, P. W., 1985. Evaluation of Regenerants for the removal of uranyl carbonates from strong-base anion exchange resin. Masters Thesis, Environmental Engineering, University of Cincinnati.

WHO, 1979. Radiological examination of drinking-water. World Health Organization, Euroreport 17.

WHO, 2004. Uranium in drinking water. Background document for the development of WHO guidelines for drinking-water quality. WHO/SDE/WSH/03.04/118.

Wolery T. J., 1992. EQ3/6, A Software Package for Geochemical Modeling of Aqueous Systems: Package Overview and Installation Guide: UCRL-MA-110662-PT-I, Lawrence Livermore National Laboratory, Livermore, California, 1992.

Zielinski R. A., Asher-Bolinder S., Meier A. L., Johnson C. A., Szabo B. J., 1997. Natural or fertilizer-derived uranium in irrigation drainage: a case study in southwestern Colorado, U.S.A. Applied Geochemistry, 12, 9-21.

In: Drinking Water and Water Management ISBN: 978-1-63117-523-7
Editor: Mason Giannino

Chapter 5

PASSIVE SAMPLING: AN INNOVATIVE APPROACH TO STUDY ENDOCRINE DISRUPTING COMPOUNDS IN WATER

Emanuele Magi*
Department of Chemistry and Industrial Chemistry,
University of Genoa, Genoa, Italy

ABSTRACT

Endocrine disrupting compounds represent a wide variety of pollutants, natural and synthetic, among which the steroidal hormones exhibit the highest estrogenic activities.

Since water monitoring for these compounds can be challenging, sensitive analytical techniques are required but also the sampling technique has to be considered. Most conventional screening methods in water involve spot sampling, providing only a snapshot of the levels of pollutants at the time of sampling. However the concentrations of pollutants can fluctuate very rapidly depending on occasional releases or environmental conditions and frequent sampling is often required. Passive sampling represent an alternative approach: it provides the average concentration of the monitored compounds in the site as a function of the exposure time, enabling also the in-situ preconcentration of target analytes.

---

* Tel: 39.010.3536187. fax: 39.010.3536190, E-mail: emanuele.magi@unige.it.

In our laboratories a fast analytical method using LC-MS/MS was developed to determine five EDCs, often included in studies regarding EDC monitoring in waters, since they are considered representative of phenolic compounds and steroid estrogens: 4-n-nonylphenol, bisphenol A, estrone, 17β-estradiol and 17α-ethinylestradiol. A case study on raw surface water and drinking water was taken into account using the approach of water passive sampling: obtained results are here presented and discussed.

INTRODUCTION

Endocrine disrupting compounds (EDCs) are emerging contaminants that can alter the normal functioning of the endocrine system in wildlife and humans. Their potential adverse effects in several aquatic organisms, have been well documented in the literature [1]. EDCs comprise a wide variety of natural and synthetic chemicals: among these, the steroidal hormones (estrogens and progestogens) exhibit the highest estrogenic activities.

In several European countries, the steroidal hormones have been measured at the ng/L level in natural waters [2]. Due to their potential adverse health effects and persistence in the aquatic environment, contamination of surface water and groundwater with EDCs has been recognized as an important issue in many countries and there is a growing interest in understanding their fate during drinking water treatment [3].

Analytical methods reported in the literature for the determination of EDCs in aqueous matrices are usually based on spot sampling (or "grab sampling") followed by a laboratory-based preconcentration (typically solid phase extraction, SPE) and subsequent analysis [4, 5]. This approach, regularly used in monitoring plans, has some limitations; in particular it provides the instantaneous measurement of pollutant levels and suffers from the uncertainty of short- and long-term concentration variations. In the framework of environmental monitoring of micropollutants in aquatic ecosystems, the use of an integrative sampling represents a very effective alternative approach.

In order to obtain a time-integrated evaluation of the bioavailability of different types of contaminants, bivalve molluscs are often used as bioindicators [6, 7]. These animals are filter-feeding organisms able to accumulate pollutants, even if they are present at extremely low concentrations; their ability in concentrating substances can be up to three orders of magnitude.

In recent years, as an alternative approach to biomonitoring, passive sampling showed a rapid development and different sampling devices for continuous monitoring of environmental contaminants have been profitably employed. Due to the dynamic interaction with high volumes of water, passive samplers can perform integrative sampling of various substances, with an exposure period that can reach 5-6 weeks [8]. Therefore, they allows the preconcentration of ultra-trace compounds and the revealing of contaminants deriving from episodic events, not always detectable by spot sampling. Moreover, they mimic the biological accumulation and offer advantages in handling, use and cost, when compared with conventional techniques of spot sampling [9]. The results represent the time integrated average concentration (TWA Time-Weighted Average), which provides basic information in the ecological risk assessment, due to chemical type stressors. The TWA data collection by spot-sampling requires frequent sampling, high costs and large number of samples to be analyzed [10].

Passive sampling is based on the spontaneous flow of an analyte from the sampled medium to the receiving phase of the sampler; this flow is generated by the difference between the chemical potential of the analyte in the two media.

The advantages of passive sampling can be summarized as follows:

- it brings together in a single step, sampling, purification (selective extraction) and *in-situ* preconcentration
- it allows the detection of episodic contaminations at ultratrace level
- it usually does not affect the speciation, sampling only the dissolved phase
- it does not require energy and it simplifies the sample preparation procedure.

Experimental and Results

In our laboratory, an analytical methodology for the determination of some of the most well-known endocrine disruptors in water (nonylphenol, bisphenol A, estrone, estradiol and ethinyl estradiol) was developed [11]. The sample collection and treatment was based on the use of passive sampling by Polar Organic Chemical Integrative Sampler (POCIS); these sampling devices were developed by Alvarez et al. in 2004 [12] to sample and preconcentrate polar organic compounds. They consists of two microporous membranes of polyethersulfone that enclose a triphasic stationary phase, the mixture Isolute ENV/Ambersorb1500/SX-3Biobeads; the analytes diffuse from water through the semipermeable membrane and are adsorbed within the phase (Figure 1).

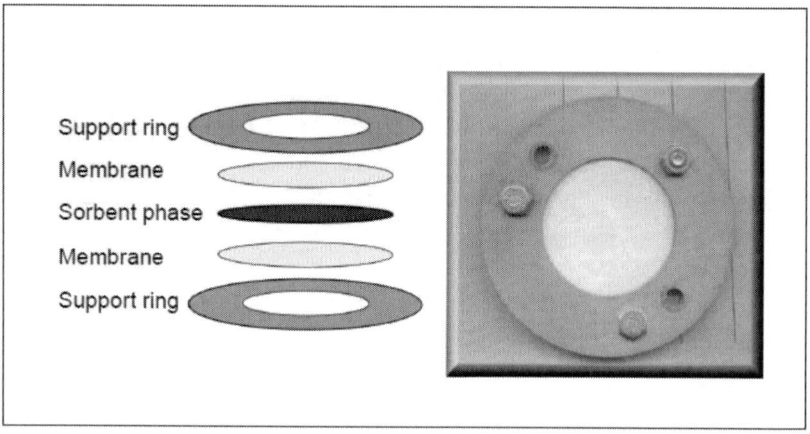

Figure 1. Description of the components of a POCIS and photograph of the sampler.

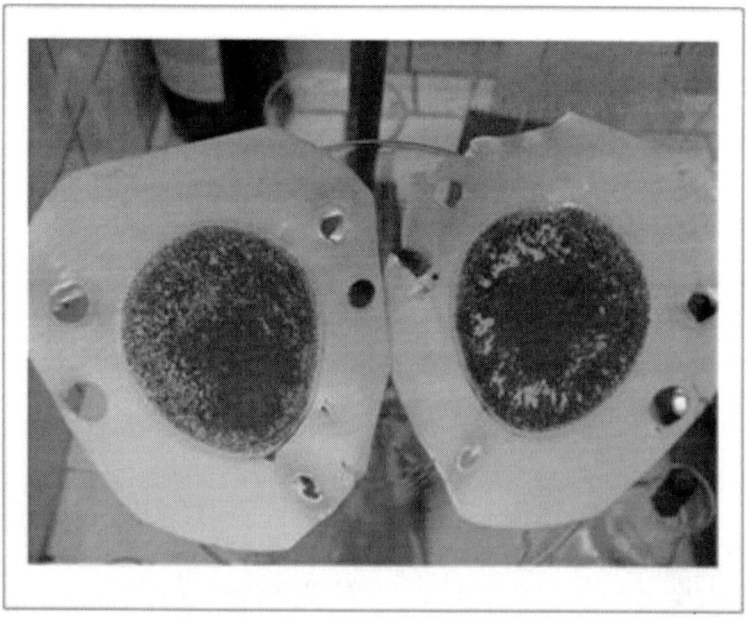

Figure 2. The two membranes of an open POCIS with the triphasic sorbent (Isolute ENV/Ambersorb1500/SX-3Biobeads).

We report here a simple case study concerning the monitoring of surface water by means of POCIS. The samplers were deployed in double at the entrance and exit of a drinking water treatment plant in Liguria (Italy), to determine the possible presence of EDCs. A couple of samplers was exposed for two weeks, another pair for four weeks, in order to verify the integrated behaviour of POCIS and exploit the ability of in situ preconcentration.

After the exposure, the samplers were opened in the laboratory (Figure 2), the sorbent phase transferred in a glass column and eluted with a tetrahydrofuran, acetone and methanol mixture.

The eluate was evaporated to dryness and redissolved in a methanol-water mixture, adding the labeled internal standard (deuterated bisphenol A); the obtained sample is analyzed by liquid chromatography - mass spectrometry tandem (LC-MS/MS). The instrumentation consisted of a Liquid Chromatograph Agilent 1200 Series SL coupled to a Triple Quadrupole Mass Spectrometer Agilent 6430 MSD with API-Electrospray source, in negative ion mode.

The chromatographic method "fast LC" developed in our labs allows the separation of the analytes in less than four minutes, thanks to the use of a column with sub-2-μm particles (Pinnacle DB Biphenylic, Restek), which provides better efficiency and resolution than conventional columns [13]. The sensitivity required for the low concentration levels expected in water samples was achieved using the triple quadrupole in the Selected Reaction Monitoring (SRM) mode, i.e., monitoring two transitions for each analyte. In Table 1 the retention times of chromatographic separation and the limits of detection and quantitation (LOD and LOQ) expressed as ng of analyte per mL of solution analyzed, are shown.

Table 1. Retention time and limits of detection and quantitation of the five EDCs

Analyte	Retention time (min)	LOD (ng/mL)	LOQ (ng/mL)
Bisphenol A	1.6	0.3	0.9
Estradiol	2.0	0.4	1.1
Ethinylestradiol	2.3	0.4	1.2
Estrone	2.5	0.3	0.8
Nonylphenol	3.8	0.4	1.3

Results obtained from the POCIS extracts deployed in drinking water (exit of water treatment plan) did not revealed any traces of analyte (overlapping blank chromatograms).

In raw water, instead, all considered steroidal compounds (estrone, estradiol and ethinyl estradiol) were below the limit of detection (LOD) or the limit of quantitation (LOQ), while nonylphenol and bisphenol A (BPA) were observed. In particular the concentrations of BPA, expressed as absolute quantity mediated on two samplers, were 46 ng/POCIS in samplers exposed for 17 days and 95 ng/POCIS in those exposed for 30 days.

This result confirms the integrative behaviour of the sampler that continues to accumulate the analyte in an approximately linear way over time. This consideration, together with the low quantity measured, seems to exclude spot contamination episodes during the sampling period.

In order to obtain the TWA values it is necessary to use the equation [5]:

$$C_w = C_s * M_s / R_s * t$$

where **Cw** and **Cs** respectively represent the concentrations of the compound in water (ng/L) and in the sorbent phase of the sampler (ng/g), **Ms** is the quantity of the sorbent phase (g), **Rs** is a kinetic constant (sampling rate, L/day) and **t** is the sampling period (days). **Cw** corresponds to the TWA and it can be calculated by having available the **Rs** value. The value of this constant changes from analyte to analyte and it can be obtained experimentally by laboratory calibrations. For some compounds it is possible to find the values of **Rs** in literature and use them to solve the above equation, thus obtaining an acceptable evaluation of the TWA.

In the specific case of BPA, we used a **Rs** value of 0.088 L/day, reported by Arditsouglou et al. [14]. The subsequent calculation of the TWA provides us the value of 30.8 ng/L, which represents the average concentration of the raw water sampled in the first two weeks; the TWA calculated for the whole period (30 days) is instead equal to 35.9 ng/L. The two values are very close, indicating a low level of contamination that remains stable in time, as already discussed.

During the same period, some spot samplings were performed at the entrance of the cosidered drinking water treatment plant and analyzed: Bisphenol A was always below the LOD of the method (50 ng/L). This is not surprising considering the TWA values obtained with POCIS (30-35 ng/L); POCIS integrative sampling of shows very well its high capacity to preconcentrate the contaminant present in the water at ultra-trace level, allowing the detection and a semi-quantitative evaluation in terms of TWA.

Similar considerations can be made for nonylphenol. Since the steroidal compounds were never detected nor by spot sampling nor integrative sampling, we can assert that TWA data are in acceptable agreement with those obtained from traditional sampling, highlighting that the considered site is not significantly affected by EDC contamination.

REFERENCES

[1] Hill M, Stabile C, Steffen K, Hill A. *Environ. Pollut.* 2002, 117, 295-300.

[2] Kuster M, Azevedo DA, López de Alda MJ, Aquino Neto FR, Barceló D *Environ. Int.* 2009, 35, 997-1003.

[3] Benotti M.J., Stanford B.D., Wert E.C., Snyder S.A. *Water Res.* 2009, 43, 1513-1522.

[4] Céspedes, R., Lacorte, S., Raldúa, D., Ginebreda, A., Barceló, D., Piña, B. *Chemosphere,* 2005, 61, 1710-1719.

[5] Kuster M., Lopez de Alda M.J., Rodriguez-Mozaz S., Barceló D. In: M Petrovic, D Barceló (Eds), *Comprehensive Analytical Chemistry,* Elsevier 2007.

[6] Magi E., Liscio C., Di Carro M. J. *Chromatogr.* A 2008, 1210, 99-107.

[7] Di Carro M., Magi E., In: Toma J. Quintin (Eds), *Chromatography Types, Techniques and Methods,* Nova Science Publishers 2010.

[8] Zhang Z., Hibberd A:, Zhou J.L. *Analytica Chimica Acta* 2008, 607, 37-44.

[9] Magi E., Di Carro M., Liscio C. *Anal. Bioanal. Chem.* 2010, 397, 1335-1345.

[10] Vrana B., Mills G.A., Allan I.J., Dominiak E., Svensson K., Knutsson J., Morrison G., Greenwood R. *Trends Anal. Chem.* 2005, 24, 845–868.

[11] Di Carro M., Scapolla C., Liscio C., Magi E. *Anal. Bioanal. Chem.,* 2010, 398, 1025–1034.

[12] Alvarez D.A., Petty J.D., Huckins J.N., Jones-Lepp T.L., Getting D.T., Goddard J.P., Manahan S.E. *Environ. Toxicol. Chem.* 2004, 23, 1640-1648.

[13] Magi E., Scapolla C., Di Carro M., Liscio C.J. *Mass Spectrom.,* 2010, 45, 1003–1011.

[14] Arditsoglou A., Voutsa D. *Environ. Pollut.* 2008, 156, 316-324.

In: Drinking Water and Water Management
Editor: Mason Giannino
ISBN: 978-1-63117-523-7

Chapter 6

PREDOMINANT BACTERIUM STRENGTHENING TECHNOLOGY FOR QUINOLINE DEGRADATION IN COKING WASTEWATER: PERFORMANCE EVALUATION AND PRACTICAL APPLICATION

Wenying Li*[*], *Jieying Jing, Mengjun Sun and Jifeng Hua

Key Laboratory of Coal Science and Technology (Taiyuan University of Technology), Ministry of Education and Shanxi Province; Training Base of State Key Laboratory of Coal Science and Technology Jointly Constructed by Shanxi Province and Ministry of Science and Technology, Taiyuan, P. R. China

ABSTRACT

The environmental persistence and fate of quinoline in coking wastewater is of particular interest because it mainly contributed to COD concentration in effluent, and made the effluent discharge not be well satisfied with sewage discharge standard (GB 13456-2012). In order to

[*] Corresponding author. Email: ying@tyut.edu.cn; Tel & fax: +86-351-6018453.

efficiently remove quinoline from coking wastewater, we employed biological strengthening technology to degrade quinoline, investigated the influence of environmental factors on the degradation performance of quinoline and its application in treating real coking wastewater. A highly efficient quinoline-degrading bacteria strain *Comamonas testosterone sp.* Y-01 was screened and isolated from activated sludge taken from coking wastewater. Biodegradation experiments showed Y-01 had strong ability for quinoline degradation, degrading as much as 100 $mg \cdot L^{-1}$ of quinoline within 12 hours and the biodegradation efficiency of quinoline reached 96-99%. Y-01 can almost completely remove quinoline in an environment with pH values between 4 and 10 at 30 °C. External inorganic-nitrogen and carbon sources apparently had no influence on the quinoline degradation ability of this strain. External phenol, however, can accelerate the biodegradation of quinoline. When this strain was mixed with phenol-degrading bacteria, Y-01's efficiency in degrading quinoline significantly increased. The predominant bacterium Y-01 was employed to replace the traditional complex microorganism on the carriers in moving-bed biofilm reactor (MBBR) to treat real coking wastewater. 74.49% quinoline is removed within 46 h, increased by 28.59-55.49% comparing to MBBR with traditional complex microorganism. Adding phenol degrading bacterium into the new dominant bacterium system in MBBR could weaken the inhibition on quinoline degrading bacterium in coking wastewater treatment, leading to a rise of 19.07% compared to without adding phenol degrading bacterium system, and 47.66-74.56% to MBBR with traditional complex microorganism. The above results will help us understand the biodegradation process and facilitate the further improvement in the practical applications of biological strengthening technology.

1. Introduction

Global legislations are putting more emphasis on the recycling of wastewater effluents, which is regarded as a strategic approach for the sustainable water management to meet the growing water demand [1, 2]. The presence of refractory organic compounds in wastewater effluent is one of the major impediments for water recycling [3, 4]. Coking wastewater is a typical industrial-heavy-polluted wastewater. It is very hard to physically or chemically degrade due to the large amounts of organic compounds produced in coal processing. With the increasingly strict wastewater emission standards, many coking plants cannot meet the requirements of both COD < 50 $mg \cdot L^{-1}$ and NH_4-N < 5 $mg \cdot L^{-1}$ (GB 13456-2012). The environmental persistence and fate of PAH (polycyclic aromatic hydrocarbon), especially quinoline in coking wastewater is of particular interest because it mainly contributed to COD concentration in effluent [5]. Besides, its variety, toxicity and persistence directly impact the health of ecosystems and present an immediate threat to human beings via contaminating drinking water supplies (i.e., surface and ground water). Therefore, it has become a challenge to achieve the effective removal of quinoline from coking wastewater effluent to minimize pollution and to enable the water reuse.

As we all know, quinoline is difficult to biodegrade by traditional biological methods as many microorganisms don't possess the enzyme system necessary to identify the quinoline structure. In addition, quinoline retards the biodegradation of other organic pollutants from the coking wastewater because it produces poisons which effect many microorganisms. However, biodegradation is, for the time being, an important means of reducing concentrations of quinoline in coking wastewater [6]. Many microorganisms have been reported to possess the ability to degrade quinoline, such as *Pseudomonas putida* [7], *Pseudomonas stutzeri* [8], *Rhodococcus sp.* [9, 10], *Bacillus sp.* [11], *Burkholderia pickettii* [12], *Comamonas sp.* [13, 14], *Desulfobacterium indolicum* [15], *Pseudomonas sp.* [16, 17] and some fungi, such as *White rot fungus* [18] have also been reported. Additionally, many of these species have been found to be highly efficient in the degradation of quinoline.

Environmental factors have great impact on the microbial degradation of organic compounds. They not only directly influence the organics' removal efficiency but also impact the degradation pathway. Therefore, the environment factors must be considered in the practical processing of wastewater. Previous works related to the quinoline biodegradation focused on the degradation rate of bacteria [19] and biodegradation pathway [20], while only few studies have centered on the effects of environmental factors and its real performance in practical application.

In order to efficiently remove quinoline from coking wastewater, biological strengthening technology was employed to degrade quinoline. A gram negative aerobe was isolated from activated sludge in the aeration tank of the Anaerobic-Anoxic-Oxic (A^2/O) process of the coking wastewater treatment system. This paper mainly focuses on the investigation about environmental factor influences on the biodegradation of quinoline by this strain and its biodegradation performance in real coking wastewater treatment.

2. Experimental

2.1. Materials and Growth Conditions

Mineral salt medium (MSM): 0.5 $g \cdot L^{-1}$ KH_2PO_4; 0.2 $g \cdot L^{-1}$ $MgSO_4 \cdot 7H_2O$; 0.02 $g \cdot L^{-1}$ $FeSO_4 \cdot 7H_2O$; 1.0 $g \cdot L^{-1}$ NaCl; 1% trace element reserve liquid; 0.1 MPa, 121 °C moist heat sterilization for 20 min. (Notes: trace element reserve liquid was prepared alone and sterilized, and it would be added to the medium when being used.) Quinoline was added to the sterilized MSM in varying concentrations as the sole carbon source and the sole nitrogen source.

Trace element reserve liquid: 2.0 $g \cdot L^{-1}$ $CaCl_2 \cdot 2H_2O$; 0.128 $g \cdot L^{-1}$ $CuSO_4 \cdot 5H_2O$; 0.2 $g \cdot L^{-1}$ $ZnSO_4 \cdot 7H_2O$; 0.152 $g \cdot L^{-1}$ $MnSO_4 \cdot H_2O$; 0.006 $g \cdot L^{-1}$ H_3BO_3; 0.1 MPa, 121 °C moist heat sterilization for 20 min. It was preserved at 4 °C after cooling.

Luria-Bertani (LB) medium: 10 $g \cdot L^{-1}$ peptone; 5 $g \cdot L^{-1}$ beef extract; 10 $g \cdot L^{-1}$ NaCl; 121 °C moist heat sterilization for 20 min. LB medium was used for bacterial enrichment and maintenance.

Solid medium: 2% agar powder is added to the medium.

2.2. Isolation and Purification of Quinoline-Degrading Strains

The activated sludge was extracted from the coking wastewater treatment plants in Taiyuan, Shanxi province, P.R. China. First, the sample of the activated sludge was placed in the LB medium for enrichment culture for 3 days at 30 °C, shaking at 180 $r \cdot min^{-1}$. Several grains of glass beads were added for the purpose of deflocculating. Second, volumes of MSM between 10 and 90 mL were combined with 100 $mg \cdot L^{-1}$ quinoline as the sole carbon and nitrogen source and shaken at 120 $r \cdot min^{-1}$ with a constant temperature of 30 °C. Finally, 5 mL mixture solution of MSM and quinoline was transferred to fresh MSM each 3 days through increasing the concentration of quinoline by 100 $mg \cdot L^{-1}$. In other words, quinoline (the concentrations of quinoline used were 100, 200, 300, 400, 500, 600 and 700 $mg \cdot L^{-1}$ sequentially) was added to the solution over a period of time in order to gradually increase the concentration until no strains were found to be growing.

Pure strains can be isolated and purified by the spread plate method and the successive streak plate method. The solid medium is MSM with 500 mg·L^{-1} quinoline. After being enriched culture by LB medium, pure strains were isolated and transferred into solutions of 80% of glycerol and either preserved at -20 °C in the refrigerator or as a slant culture at 4 °C.

The phenol-degrading bacteria used in this study are isolated in our lab. More detailed information can be found in our previous work [21].

2.3. Strains Identify by 16S rDNA Gene Sequence Analysis

By 16S rDNA gene sequence analysis (Genolab Biotech Co. Ltd.), sequencing results were examined for homology contrasted by the Blast of the National Center of Biotechnology Information (NCBI). Homologous sequences were found in GenBank, and then sequences edited by Bioedit and MEGA were used to build system growth trees using the Neighbor-Joining (NJ) method [22].

2.4. Preparation of Suspended Bacteria Liquid

Inoculated strains were placed in LB medium (500 mg·L^{-1} quinoline was added to prevent the inoculated strains from other bacteria) and cultured for 24 hours with 120 r·min^{-1} shaking at 30 °C, and then centrifuged at 10000 r·min^{-1} for 10 min. The bacteria were then collected and washed 3 times with MSM, and finally the bacteria was suspended in MSM to make the suspended bacteria liquid (OD_{600} = 1).

2.5. Biodegradation of Quinoline and Analysis

10 vol% inoculated suspended bacteria liquid was added to MSM of varying initial quinoline concentrations. Cells were grown at 30 °C with shaking at 120 r·min^{-1}, sampled at every 2 hours intervals. Portions of the samples were used to assay cell densities and the analysis and investigation of quinoline. Experiments were performed at varying initial quinoline concentrations, temperature and pH to investigate the degradation performance and optimal degradation conditions of the bacteria. Experiments were performed with external carbon source and nitrogen source, as well as a

mixture of strain and phenol degrading bacteria in order to investigate the degradation performance of extrinsic factors. The same MSM sample without bacteria was used as a blank control throughout the whole process.

Cell density was monitored by measuring the absorbance at a wavelength of 600 nm in a UV spectrophotometer.

To measure the concentration of no degradable quinoline, some culture solutions were placed in the centrifuge at 10000 $r \cdot min^{-1}$ for 10 min, cell supernatant was taken out and extracted by ethyl acetate, and then was monitored by spectrophotometer at a wavelength of 313 nm. Eventually, the concentration of quinoline was worked out by the standard curve. The degradation rate of quinoline can be expressed as follow [23].

$$\eta = \frac{C_0 - C_1}{C_0} \times 100\%$$

Where: η — degradation rate of quinoline, %;

C_0— initial concentration of quinoline, $mg \cdot L^{-1}$;

C_1— the concentration of quinoline degraded, $mg \cdot L^{-1}$;

3. Results and Discussion

3.1. Properties of the Isolated Strain

According to our experiments, four pure bacterial strains were found to have the ability to utilize quinoline as the sole carbon, nitrogen and energy source from the activated sludge. By surveying the strains' growth and degradation of quinoline in different initial quinoline concentrations, one highly efficient quinoline-degrading strain with strong endurance capacity of quinoline was isolated, Y-01, which is a gram negative aerobe. When Y-01 degraded quinoline, the culture solution initially turned pink and then finally became a light, pale yellow. Some researchers have observed many kinds of color changes [24, 25]. It is indicated that the microbial degradation of quinoline exists diversely and different strains degrade quinoline into different intermediates.

16S rDNA gene fragments (1409 bp) of Y-01 were entered into the GenBank and their homology contrasted using the Blast of the National Center of Biotechnology Information (NCBI). The homologous sequence was found in the GenBank, and then the system growth trees were built using the

Neighbor-Joining (NJ) method. The results as shown in Figure 1 indicated that the 16S rDNA gene fragments of Y-01 have a greater than 99% homology with *Comamonas testosterone* (AM184216), and those strains of makeup similar to that of Y-01 all belong to *Comamonas testosterone sp.* Therefore, it is confirmed that Y-01 is *Comamonas testosterone.*

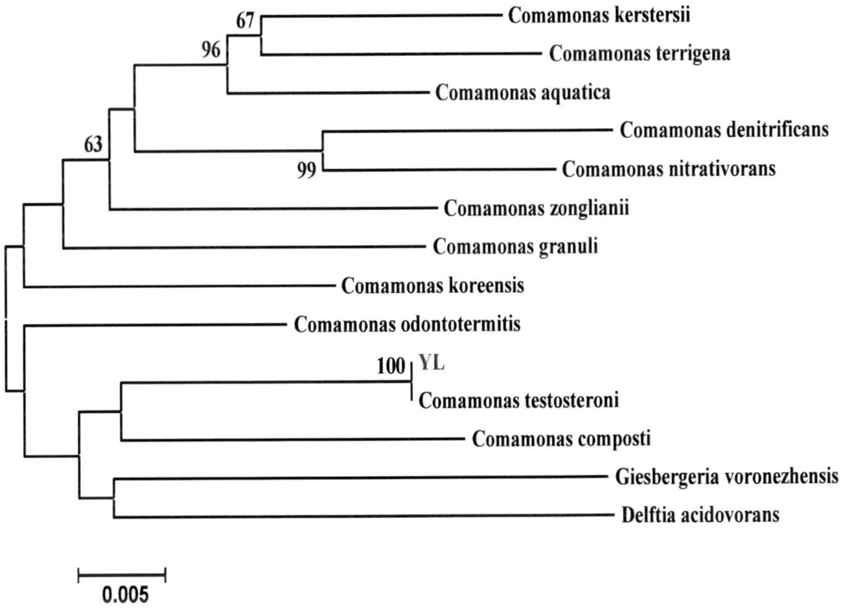

Figure 1. Phylogenetic tree analysis based on the partial 16S rDNA.

3.2. Environmental Factor Influence on Quinoline Biodegradation

Environmental factors can greatly affect the microbial degradation efficiency of quinoline. It is therefore important to learn the significant environmental factors influencing the biodegradation of quinoline; this eventually helps us understand the biodegradation process and facilities the further improvement in the practical applications of biological strengthening technology.

3.2.1. pH

The biodegradation efficiency of quinoline by Y-01 was investigated at a 100 mg·L^{-1} initial concentration of quinoline, 10% inoculation, 120 r·min^{-1} and pH 4, 6, 7, 8, 10 at 30 °C.

From Figure 2, it was found that the degradation efficiency of quinoline could reach 96 -99 % within 12 hours in the range of pH 4 - 10, however a more alkaline environment gave slightly better results. This seems to indicate that this strain could adapt to a wide pH range. The acidity of the solution not only had an influence on the bacterial activity, but also on the resulting form of the substrate. In water solution, quinoline balanced as follows (pKa = 4.8) [26]:

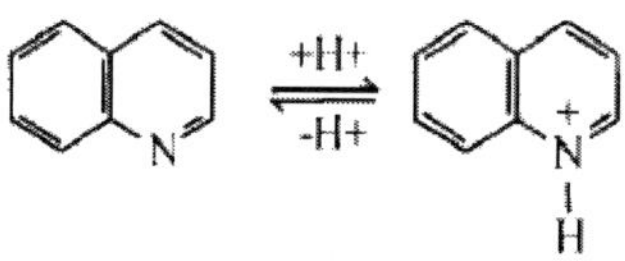

In acidic conditions, quinoline existed in the form with a positively charged proton; while in alkaline conditions, quinoline took the form of a neutral molecule which was more easily degraded by bacteria. So a partially alkaline system would be more beneficial to the degradation of quinoline.

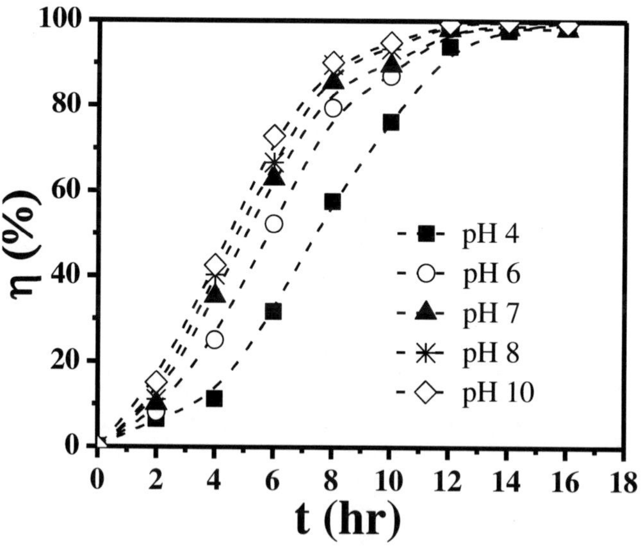

Figure 2. Effect of pH value on the biodegradation efficiency of quinoline.

3.2.2. Temperature

To examine temperature influence in the quinoline degradation process, the degradation efficiency of quinoline by Y-01 was studied under the following conditions: an initial concentration of 100 $mg \cdot L^{-1}$ quinoline, 10% inoculation, pH 7 and 120 $r \cdot min^{-1}$ rotation speed of solution. The test temperatures were carried out at 10, 20, 30 and 40 °C separately. It is suggested in Figure 3 that temperature had a great influence on the degradation efficiency of quinoline. For example, the degradation efficiency of quinoline was only reached 20% in 12 hours at 10 °C. As temperature increased, the quinoline degradation efficiency increased significantly. However, we should note that the quinoline degradation efficiency at 40 °C was lower than that at 20 °C and 30 °C.

In the microbial degradation of organic pollutants, the enzymatic reaction plays an important role. There are always two sides to everything, including the influence of temperature on the enzyme reaction. On the one hand, when the temperature increases, the molecular movement is accelerated, correspondingly, the opportunity for intermolecular collisions is higher, resulting in an increase in the reaction rate. On the other hand, as a biological catalyst, the chemical essence of enzymes is protein. High temperatures potentially lead to an increase in enzyme denaturation and thereby slow down the reaction process. The optimum temperature of enzymatic reaction is at the equilibrium point of these two processes in a system. It was found that the optimum temperature of Y-01 in this experiment was 30 °C. When the temperature was below or above this point, either the growth of bacteria or the enzyme activity was inhibited, and therefore the degradation efficiency of the quinoline became low.

3.2.3. Initial Quinoline Concentration

The influence of different initial quinoline concentrations on quinoline degradation efficiency by *Comamonas testosterone sp.* Y-01 was investigated. The test conditions were at pH 7, 10% inoculation, 30 °C, 120 $r \cdot min^{-1}$ rotation speed of solution. The initial concentration examined were 0, 100, 150, 350, 500 or 700 $mg \cdot L^{-1}$ of quinoline.

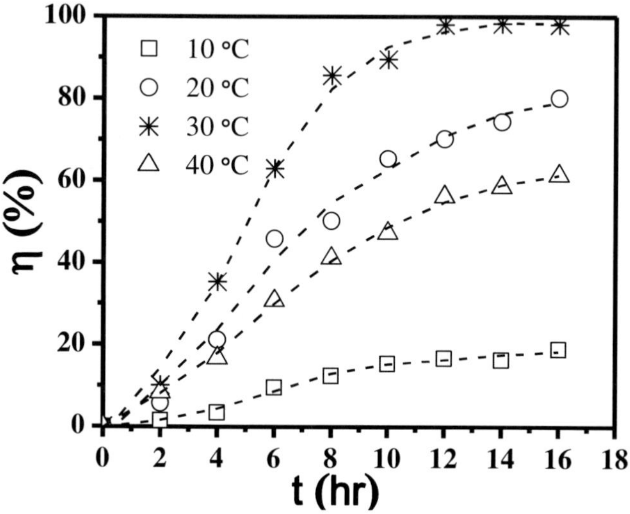

Figure 3. Effect of temperature on the biodegradation efficiency of quinoline.

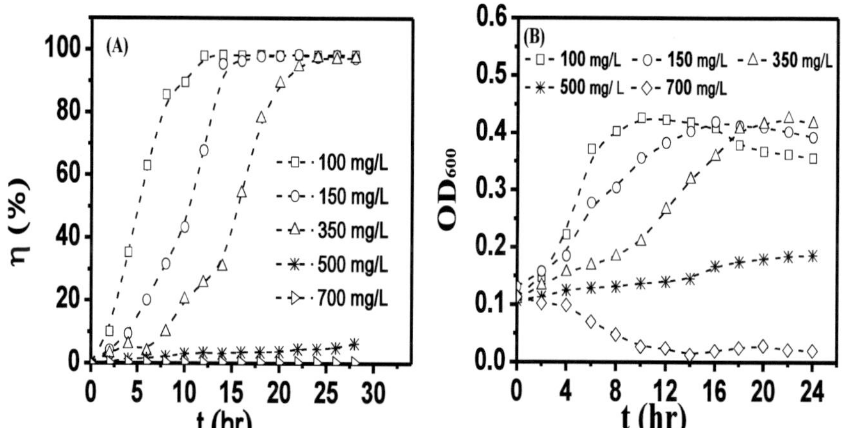

Figure 4. Effect of initial concentration on the biodegradation efficiency of quinoline (A) and its influence on the growth of *Comamonas testosterone* Y-01 strain (B).

Degradation efficiency of quinoline and the growth curves of *Comamonas testosterone* Y-01 strain at different initial quinoline concentrations were presented in Figure 4. There was no obvious quinoline degradation in the control solution. Figure 4(A) showed that as the initial concentration of quinoline increased, the biodegradation efficiency obviously decreased. When the initial quinoline concentration is 700 mg·L^{-1}, it became toxic and Y-01 could not survive (Figure 4(B)). The results indicated that increasing initial quinoline concentrations inhibited the activity of the *Comamonas testosterone* Y-01, but it could still survive and keep degradation activity in a wide range of concentrations. It was shown that this strain could degrade a solution of 100 $mg \cdot L^{-1}$ of quinoline within 12 hours and 350 $mg \cdot L^{-1}$ of quinoline within 24 hours, and correspondingly the removal efficiencies of quinoline at these concentrations could reach 96-99%. This further supported the hypothesis that the *Comamonas testosterone* Y-01 bacterium is a highly efficient quinoline-degrading strain.

3.2.4. External Inorganic-Nitrogen, Carbon and Phenol Source

Complex pollutants (such as polycyclic aromatic hydrocarbons) and phenolic compounds are the main pollutants in coking wastewater. In order to analyze the effect of phenolic compounds such as external inorganic-nitrogen, carbon and phenol source on the degradation of quinoline, the degradation efficiency of quinoline by *Comamonas testosterone sp.* Y-01 was performed under the following conditions: an initial concentration of 100 $mg \cdot L^{-1}$ quinoline, pH 7, 10% inoculation, 30 °C, and 120 $r \cdot min^{-1}$ rotation speed of solution. The external inorganic-nitrogen, carbon and phenol source was provided by NH_4NO_3, glucose and phenol.

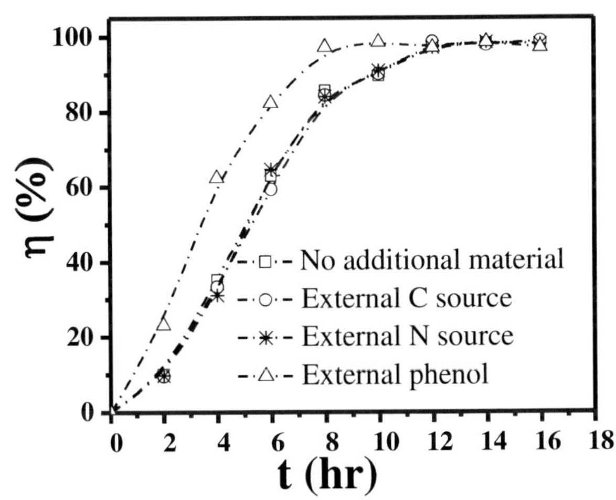

Figure 5. Effect of external inorganic-N, carbon and phenol source on the biodegradation efficiency of quinoline.

It was seen from Figure 5 that the presence of an external inorganic-nitrogen and carbon source had no effect on the quinoline degradation.

The nitrogen and carbon sources Y-01 needed mainly came from quinoline itself. But the presence of external phenol significantly assisted to accelerate the biodegradation of quinoline. Furthermore, it was possible that the presence of other organics might interfere with the biodegradation of quinoline. Therefore, adding Y-01 to coking wastewater with abundant phenol would be of practical significance.

3.2.5. Phenol Degrading Bacteria

Coking wastewater contains many kinds of microorganisms, of which phenol-degrading bacteria are predominant. The influence of phenol-degrading bacteria on the quinoline degradation efficiency was studied under the same conditions as previously stated. Figure 6 showed that phenol-degrading bacteria isolated and selected from the activated sludge didn't appear to possess the function of quinoline degradation, but when Y-01 interacted with the phenol-degrading bacteria strains already presented, the degradation efficiency of quinoline raised obviously. Compared with the degradation efficiency of a single Y-01 strain, mixed strains seemed to have better degradation efficiency. Because the degradation of organic compounds by a single microbe usually produce toxic final products which could inhibit the growth of microorganisms, a variety of microorganisms constitutes a stable microbial community able to eliminate the influence of toxic degrading metabolites on bacteria [27] in the biodegradation process of organics.

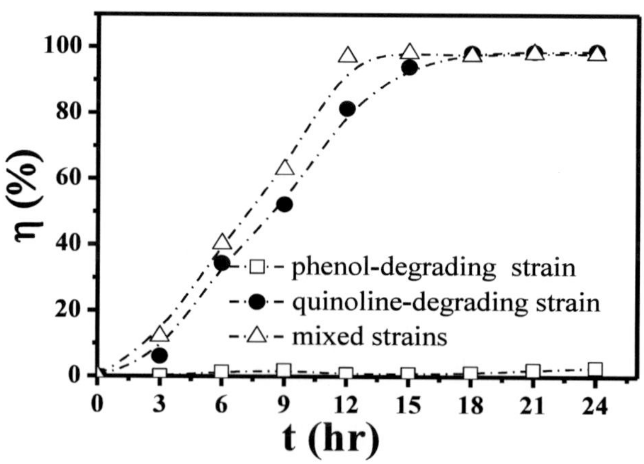

Figure 6. Influence of external phenol degrading bacteria on the biodegradation efficiency of quinoline.

3.3. Coking Wastewater Treatment by Bacterium Y-01 in MBBR

Our research group has successfully developed moving-bed biofilm reactor (MBBR) to treat coking wastewater [28, 29]. However, due to the weak competitiveness and low activity of microorganism degrading quinoline in MBBR, the quinoline degradation efficiency is only around 21.1 - 46.9%. Herein, we employed bacterium Y-01 to substitute the traditional complex microorganism on the carriers in MBBR to treat coking wastewater.

3.3.1. Quinoline Degradation by Bacterium Y-01 in MBBR

In most of coking plants, quinoline concentration in coking wastewater is less than 200 $mg{\cdot}L^{-1}$, therefore we choose 200 $mg{\cdot}L^{-1}$ as the initial quinoline concentration to investigate the degradation efficiency of quinoline in the presence of *bacterium Y-01*. Result in Figure 7 shows that the biodegradation efficiency of quinoline arrived 96.3% in 23 h by bacterium Y-01 loaded on carriers in MBBR, which has a 51.21-78.11% rise in quinoline degradation compared to MBBR with traditional complex microorganism. It indicated that bacterium Y-01 greatly accelerated the biodegradation of quinoline in MBBR, and it is an effective way to improve the efficiency of MBBR in treating wastewater.

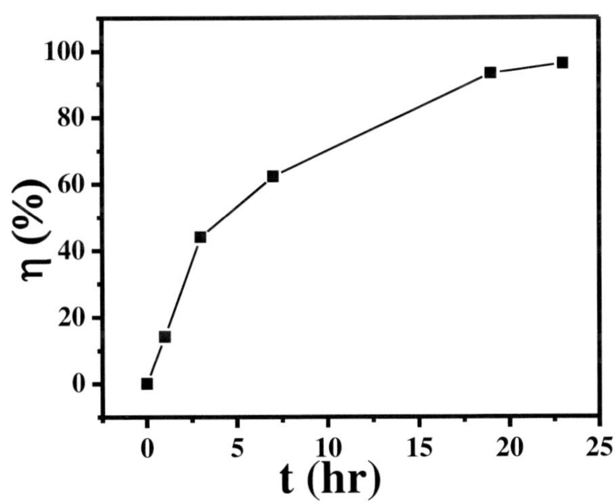

Figure 7. Degradation efficiency of quinoline in the presence of bacterium Y-01.

3.3.2. Real Cooking Wastewater by Bacterium Y-01 in MBBR

Due to the various organic compounds and all kinds of bacteria existing in real cooking wastewater, quinoline degradation is more complex and its degradation efficiency is affected by the above factors. Figure 8 presents the quinoline degradation efficiency in real cooking wastewater by Y-01 and mixture of Y-01 and phenol-degrading bacteria loaded on carriers in MBBR respectively.

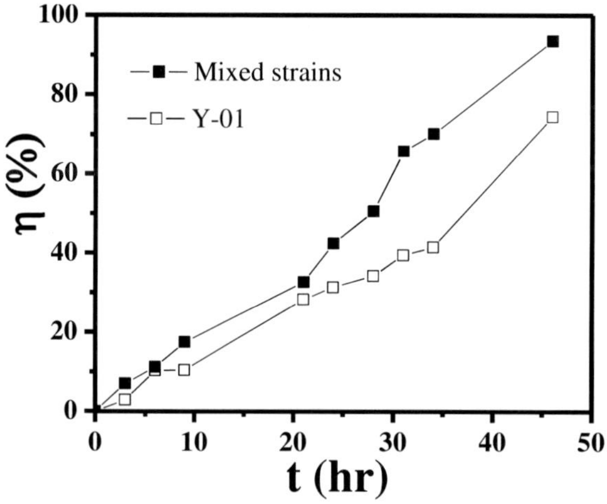

Figure 8. Biodegradation of quinoline in cooking wastewater by Y-01 and mixed bacteria (Y01 + phenol-degrading bacteria) loaded on carriers in MBBR.

It is clear that when Y-01 loaded on carriers in MBBR, 74.49% quinoline removed within 46 h, increased by 28.59-55.49% comparing to MBBR with traditional complex microorganism. After adding phenol-degrading bacteria into the new dominant bacterium system in MBBR, the degradation efficiency of quinoline increased to 93.56%, which is higher than that of Y-01. In the mixed bacteria system, the biodegradation activity of Y-01 is affected by the toxic impact from other organic compounds existing in coking wastewater and the competitive effect from phenol-degrading bacteria. While in sole Y-01 system, the inhibition effect only comes from the toxic impact of other organic compounds existing in coking wastewater. Figure 8 indicated that in the initial degradation stage, the competitive effect from phenol-degrading bacteria can be ignored, and the toxic impact plays a key role. As for sole Y-01 system, there is no change in organic compounds concentration; while in mixed bacteria system, as phenol was degraded by phenol-degrading bacteria, the phenol concentration in coking wastewater became lower, which could weaken the inhibition on quinoline degrading bacteria Y-01 in coking wastewater treatment, leading to a rise of 19.07% compared to without adding phenol degrading bacterium system, and 47.66-74.56% to MBBR with traditional complex microorganism.

CONCLUSION

A highly efficient quinoline-degrading bacteria strain *Comamonas testosterone sp.* Y-01 was screened and isolated from activated sludge taken from coking wastewater. Biodegradation experiments showed Y-01 had strong ability for quinoline degradation, degrading as much as 100 $mg \cdot L^{-1}$ of quinoline within 12 hours and the biodegradation efficiency of quinoline reached 96-99%. Y-01 can almost completely remove quinoline in an environment with pH values between 4 and 10 at 30 °C. External inorganic-nitrogen and carbon sources apparently had no influence on the quinoline degradation ability of this strain. External phenol, however, can accelerate the biodegradation of quinoline. When this strain was mixed with phenol-degrading bacteria lacking quinoline-degrading function, Y-01's efficiency in degrading quinoline significantly increased. The predominant bacterium Y-01 was employed to replace the traditional complex microorganism on the carriers in MBBR to treat real coking wastewater. 74.49% quinoline is removed within 46 h, increased by 28.59-55.49% comparing to MBBR with traditional complex microorganism. Adding phenol degrading bacterium into the new dominant bacterium system in MBBR could weaken the inhibition on quinoline degrading bacterium in coking wastewater treatment, leading to a rise of 19.07% compared to without adding phenol degrading bacterium system, and 47.66-74.56% to MBBR with traditional complex microorganism.

ACKNOWLEDGMENT

We appreciate the financial supports from the Program for Changjiang Scholars (2009), the Innovative Research Team in University (IRT0517) of MOE, and the Qualified Personnel Foundation of Taiyuan University of Technology (No: tyutrc-201301a).

REFERENCES

[1] Fujishima, A; Zhang, X; Tryk, DA. Heterogeneous photocatalysis: From water photolysis to applications in environmental cleanup [J]. *International Journal of Hydrogen Energy*, 2007, 32(4), 2664-2672.

[2] Shannon, MA; Bohn, PW; Elimelech, M; et al. Science and technology for water purification in the coming decades [J]. *Nature*, 2008, 452, 301-310.

[3] Gaya, UI; Abdullah, AH. Heterogeneous photocatalytic degradation of organic contaminants over titanium dioxide: A review of fundamentals, progress and problems [J]. *Journal of Photochemistry and Photobiology C: Photochemistry Reviews*, 2008, 9(1), 1-12.

[4] Xu, YL; Zhong, DJ; Jia, JP; et al. Enhanced dye wastewater degradation efficiency using a flowing aqueous film photoelectrocatalytic reactor [J]. *Journal of Environmental Science and Health - Part A Toxic/Hazardous Substances and Environmental Engineering*, 2008, 43(10), 1215-1222.

[5] Wu, GM; Liu, HJ; Lu, XH. COD analysis of coking wastewater treatment with activated sludge [J]. *Environment Science and Technology*, 2008, 31(1), 62-64.

[6] Rothenburger, S; Atlas, RM. Hydroxylation and biodegradation of 6-methylquinoline by pseudomonads in aqueous and non-aqueous immobilized-cell bioreactors [J]. *Applied and Environmental Microbiology*, 1993, 59(7), 2139-2144.

[7] Qiao, L; Wang, JL. Biodegradation characteristics of quinoline by pseudomonas putida [J]. *Bioresource Technology*, 2010, 101(19), 7683-7686.

[8] Shukla, OP. Microbiological degradation of quinoline by pseudomonas stutzeri: The coumarin pathway of quinoline catabolism [J]. *Microbiosource*, 1989, 59(238), 47-63.

[9] O'loughlin, EJ; Kehrmeyer, SR; Sims, GK. Isolation, characterization and substrate utilization of a quinoline-degrading bacterium [J]. *International Biodeterioration & Biodegradation*, 1996, 38(2), 107-118.

[10] Zhu, SN; Liu, DQ; Fan, L; et al. Isolation, identification and degradation characteristics of a quinoline-degrading bacterium rhodococcus sp. QL2 [J]. *Environmental Science*, 2008, 29(2), 488-493.

[11] Tuo, BH; Yan, JB; Fan, BA; et al. Biodegradation characteristics and bioaugmentation potential of a novel quinoline-degrading strain of bacillus sp. Isolated from petroleum-contaminated soil [J]. *Bioresource Technology*, 2012, 107, 55-60.

[12] Wang, JL; Wu, WZ; Zhao, X. Microbial degradation of quinoline: Kinetics study with Burkholderia pickttii [J]. *Biomedical and Environmental Sciences*, 2004, 17(1), 21-26.

[13] Cui, MC; Chen, FZ; Fu, JM. Microbial metabolism of quinoline by comamonas sp. [J]. *World Journal of Microbiology and Biotechnology*, 2004, 20(6), 539-543.

[14] Cui, MC; Li, L; Chen, FZ; et al. Study of biodegradation pathway of quinoline by comamonas testosteroni [J]. *Acta Scientiae Circumstantiae*, 2004, 24(1), 171-173.

[15] Licht, D; Johansen, SS; Arvin, E; et al. Transformation of indole and quinoline by desulfo bacterium indolicum [J]. *Applied Microbiology and Biotechnology*, 1997, 47(2), 167-172.

[16] Mohan, SV; Sistla, S; Guru, RK; et al. Microbial degradation of pyridine using pseudomonas sp. and isolation of plasmid responsible for degradation [J]. *Waste Management*, 2003, 23(2), 167-171.

[17] Sun, Q; Bai, Y; Zhao, C; et al. Aerobic biodegradation characteristics and metabolic products of quinoline by a pseudomonas strain [J]. *Bioresource Technology*, 2009, 100(21), 5030-5036.

[18] Ren, DJ; Zhang, XY. Study on degradation of quinoline in coking plant wastewater using white rot fungi [J]. *Environmental Protection Science*, 2006, 32(1), 20-23.

[19] Fetzner S. Bacterial degradation of pyridine, indole, quinoline, and their derivatives under different redox conditions [J]. *Applied Microbiology and Biotechnology*, 1998, 49(3), 237-250.

[20] Schwartz, G; Bauder, R; Speer, M; et al. Microbial metabolism of quinoline and related compounds. II. Degradation of quinoline by pseudomonas fluorescens 3, Pseudomonas putida 86 and Rhodococcus spec. B1 [J]. *Biological Chemistry Hoppe-Seyler*, 1989, 370(11), 1183-1189.

[21] Chen, C; Li, WY; Wu, JW; et al. Screening and characterization of phenol degrading bacteria for the coking wastewater treatment [J]. *Environmental Science*, 2012, 33(5), 1652-1656.

[22] Tamura, K; Dudley, J; Nei, M; et al. MEGA4: Molecular evolutionary genetics analysis (MEGA) software version 4.0 [J]. *Molecular Biology and Evolution*, 2007, 24, 1596-1599.

[23] Zhang, XX; Wang, JC; Wu, WL; et al. Isolation and identification of a strain for quinoline degradation [J]. *Journal of Agro-Environment Science*, 2008, 27(3), 959-963.

[24] Aislabie, J; Bej, AK; Hurst, H; et al. Microbial degradation of quinoline and methylquinolines [J]. *Applied Environment Microbiology*, 1990, 56(2), 345-351.

[25] Shukla, OP. Microbial transformation of quinoline by a pseudomonas sp. [J]. *Applied and Environmental Microbiology*, 1986, 51(6), 1332-1342.

[26] Nedoloujko, A; Kiwi, J. Parameters affecting the homogeneous and heterogeneous degradation of quinoline solutions in light - activated process [J]. *Journal of Photochemistry and Photobiology A: Chemistry*, 1997, 110(2), 149-157.

[27] Jin, ZG; Zhang, T; Zhu, HL; eds. Biodegradation of pollutants. 1997, East China University of Science and Technology Press. Shanghai, China.

[28] Jing, JY; Li, WY. Research status and prospect of wastewater treatment by moving bed biofilm reactor [J]. *Coal Chemical Industry*, 2008, 36(4), 23-27.

[29] Jing, JY; Feng, J; Li, WY; et al. Removal of COD from coking-plant wastewater in the moving-bed biofilm sequencing batch reactor [J]. *Korean Journal of Chemical Engineering*, 2009, 26(2), 564-568.

In: Drinking Water and Water Management ISBN: 978-1-63117-523-7
Editor: Mason Giannino

Chapter 7

IRRIGATION AGRICULTURE: SUSTAINABILITY THROUGH HOLISTIC APPROACHES TO WATER USE AND INNOVATION

Kathleen H. Bowmer[1]* and Wayne S. Meyer[2]

[1]Institute for Land, Water and Society and School of Agricultural and Wine Sciences, Charles Sturt University, Wagga Wagga, NSW, Australia

Environment Institute, University of Adelaide, Australia

ABSTRACT

Sustainable irrigation agriculture is required to provide much-needed food and fibre but is threatened by reduced water supply and increasing competition for environmental needs and urban and industrial water demand. Policy on water allocation often pits the needs of irrigation against competitive demands. This chapter explores innovative options for avoiding or mitigating this conflict. Examples include: the strategic location of forestry to minimise interception of water in the landscape; investment in infrastructure to provide environmental benefits with minimal use of more water; reducing water loss by optimising river operations; use of water sharing agreements to provide greater certainty

---

* Corresponding author: Email: kbowmer@csu.edu.au.

to irrigators and investors; market-based trading arrangements to provide flexibility to irrigators to deal with water scarcity and climate variability; improved water use efficiency both on farm and in water supply and delivery systems; and more efficient water use and re-use in cities and industry. The water cycle and water auditing provide the biophysical context for developing options by governments, policy makers and communities as they adapt to change. Processes of public participation remain a challenge but good progress is being made.

1. INTRODUCTION

Sustainable irrigation agriculture is required to provide much-needed food and fibre. The remarkably consistent ratio of 22 people per irrigated hectare in the world since 1950 is testament to the critical role that irrigation production has had in providing for the growing population. With projected population growth, the amount of water currently used in agriculture will need to double in the first half of this century to avoid widespread famine (Chartres & Varma, 2010). As well, water for sanitation, clean drinking water, urbanisation and growing dietary changes are increasing concerns. On a global scale 70 percent of available fresh water is used for agricultural production, 22% for industrial purposes and 8% for domestic purposes (Prathapar, 2000). Climate change (IPCC, 2013), and increasing competition for water for domestic and industrial purposes are likely to reduce the proportion of water available for agriculture. Thus, water scarcity is being increasingly accepted as a major limitation for increased agricultural production and food security in the 21st century (Cribb, 2010). Solutions require that scientific and hydrological understanding is combined with economic, social, institutional, ethical and legal dimensions of water management. A recent book on water resources, planning and management provides this comprehensive and interdisciplinary focus with contributions from 19 countries (Quentin Grafton & Hussey, 2011).

Sustainable water management in Australia has been reviewed recently by the Australian Academy of Technical Sciences and Engineering in a report that '*explores a framework ...that is able to adapt to future challenges through fostering the principles of improved productivity and economic prosperity as well as improved environmental and social outcomes*' (Spies & Dandy, 2012). Pressure on the use of water for food and export of virtual water are special challenges for regional Australia requiring increased effort in irrigation water use productivity (Chartres & Williams, 2006). Climate change (CSIRO, 2010), land-based interception (less run-off) through increasing afforestation (SKM et al., 2010), and competition for water for urban and industrial uses (Marsden & Pickering, 2006) threaten water availability for irrigation. Also, especially in the regulated Murray River system, there is increasing concern about impacts of irrigation on aquatic ecology and river health (Davies et al., 2010; Schofield, 2010) and, in common with the Yellow, Orange and Colorado Rivers, on downstream communities and estuarine systems (Quentin Grafton et al., 2013).

In Australia, these environmental concerns are reflected in national water policy that is largely driven by the need to address over-allocation and environmental stress in the Murray-Darling Basin. The Living Murray Program, established by the Australian government in 2002, had an initial objective to recover 500GL of water annually for the environment to be used at six environmentally important sites in the Basin. Since the mid 90's Australian Federal governments of different political persuasions have supported very significant plans to improve water management and retain more water within the ecologically stressed rivers of the Murray-Darling Basin. The most recent plan commits AUD 3.1 billion over ten years to buy-back water from irrigators and AUD 5.8 billion to invest in infrastructure to improve water use efficiency and return a share of water to the environment. The water quantity acquired by the Federal government is to be managed for environmental purposes by an independent authority, the Commonwealth Environmental Water Holder (CEWH, 2013). This reduction of water available for irrigation and its use for environmental benefit is a major challenge for Australian policy makers, irrigators and their communities. Substantial investment has been made to improve water control and distribution systems and so reduce water loss. This has decreased the need for buying water entitlements from irrigators (MDBA, 2010a) but the approach is controversial because it is expensive (Crase et al., 2013). The Council of Australian Governments recently agreed to plans to reduce overallocation in the Murray-Darling Basin by 2024 (COAG , 2013), and the full range of mechanisms used to provide sustainable levels of extraction for urban and rural water is provided by the National Water Commission (NWC, 2014).

The nexus between water and energy is an emerging issue. Proust et al. (2007) advocate closer attention to the links between climate, energy and water, noting that storage, treatment, distribution and use of water and treatment of wastewater all require energy, that energy production and use also requires water, and that climate determines demand for both energy and water. Energy is critical for access to new water sources such as desalination, for pumping from distant catchments, and for access to groundwater. Energy costs for pumping, use of farm machinery, fertilisers and pesticides are also limitations to food availability (Khan et al., 2009). While this chapter does not consider the water and energy nexus further it discusses the options for sustainable irrigation in the full knowledge that the ready availability of affordable energy will be an ultimate determinate of large scale, sustainable irrigation.

Salinity has been a major concern in Australia in previous decades (Khan et al., 2006) but a decade of drought and improvements in irrigation practice through decision support systems such as SWAGMAN (Prathapar et al., 1997) have substantially reduced the threat. As part of the Murray-Darling Basin Authority's Salinity Management Strategy 18 salt interception schemes diverted nearly 323 000 tonnes of salt from the River Murray in 2012-2013 (MDBA, 2013). Pesticide contamination of drainage water and rivers has also been substantially reduced in the last two decades (Bowmer, 2013). Though important issues for irrigation sustainability, salinity and water quality are not considered further in this chapter except where they are strongly linked with water use.

Options for amelioration of threats to irrigation are described in this chapter through a 'big picture' view of the water cycle. The water cycle and water auditing provide the biophysical context for developing options so that governments, policy makers and communities can work together as they adapt to change.

Examples of innovation are detailed in the following text. These include: attention to water supply in the catchment including the strategic location of forestry to minimise interception of surface run-off water; options to benefit the environment with minimal impact on irrigation water availability; water savings through investment in infrastructure; improved efficiency of water delivery and river operations; use of water sharing agreements to provide greater certainty to irrigators and investors; introduction of market-based trading arrangements in which the environment is a statutory water holder; increased water use efficiency on farm; and reduction in pressure on water supplies through more efficient water use and re-use in industry and cities. Finally, procedural fairness and transparent decision making is required to build confidence in reform and to capture and implement innovative ideas and new markets. While this remains a challenge for water planners in Australia, considerable progress has been made.

While this chapter focuses on Australia, many of the threats and options for amelioration will be applicable in an international context.

2. IRRIGATION IN AUSTRALIA

The use of water for irrigation (Tables 1 and 2) shows the dominance of the Murray-Darling Basin (MDB) which produces over one-third of the nation's food supply (Figure1). The MDB contains over two million people and 65% of Australia's irrigated land. It also supplies approximately half of the major city of Adelaide's water requirements via pipeline connections to the Murray River. Irrigation is also important in the Ord Irrigation Scheme in the north of Western Australia, and in the Burdekin, Queensland (NLWRA, 2002). Other new developments in Tasmania and the north of Australia are progressing although with some controversy (CSIRO, 2009; Denholm, 2010). Irrigation in the MDB was reviewed by the Cooperative Research Centre for Irrigation Futures (Meyer et al., 2005; MDBA, 2010b). In the MDB rainfall is highly variable both within and between years, and many dams and weirs, with a total storage of about 35 000 GL, have been built over the last century to smooth water supply.

Table 1. Irrigation water use on Australian farms 2008-9 by enterprise (ABS, 2010)

Crop	Water use in MDB (GL)	Water use external to MDB (GL)	Total water use (GL)
Pasture	760	1 221	1 981
Cereals	789	1 552	2 341
Rice	101	0	101
Sugar cane	0	761	761
Cotton	793	86	879
Other broad-acre	90	55	145
Fruit, nuts, berries	374	223	597
Vegetables	122	312	434
Grapevines	439	57	496
Total	3 492	3 008	6 500

Table 2. Irrigation water use on Australian farms 2008-9 (ABS, 2010)

	In MDB	External to MDB	Total
Area watered with irrigation (ha 000s)	929	832	1 761
Volume applied (GL)	3 492	3 008	6 500
Av application rate (ML/ha)	3.8	3.6	-
Supplied by government or private irrigation schemes (GL)	1 573	1 032	2 605
Surface water used (GL)	1 032	943	1 975
Groundwater used (GL)	1 068	1 421	2 489
Recycled and re-used from off- farm (GL)	41	69	110
Surface irrigation (ha 000s)	596	207	803
Drip or trickle above ground (ha 000s)	115	109	224
Drip or trickle sub-surface (ha 000s)	13	13	26
Microspray (ha 000s)	34	51	85
Other sprinkler (including portable, hose, large mobile machines) (ha 000s)	155	443	598
Other irrigation systems (ha 000s)	43	52	95

Rights to access water are commonly divided into two categories: *water entitlements (also known as 'permanent water', with* a long-term access to a share of the total consumptive pool of the water resource) and water *allocations (also known as 'temporary water').* The volume or share of water that water entitlement holders receive during a given water year is dependent on the available water in storages, expected inflows, system losses, demand expectations and delivery capacity (NWC, 2011b).

Cullen (2011) anticipated that irrigators in Australia will double the wealth obtained from around half the water previously available. He predicted: a reduction in permanent planting with more emphasis on annual crops that can be planted once water availability for the season is known; larger properties able to grow both perennial and opportunistic crops; an overall contraction in the area irrigated leading to stranded assets and increasing costs for those remaining; an increase in the value of water which will continue to drive improvements in water use productivity; a departure of dairy farming from irrigation areas with purchase of fodder; and new irrigation technologies that may see crops move from heavier to lighter-textured soils.

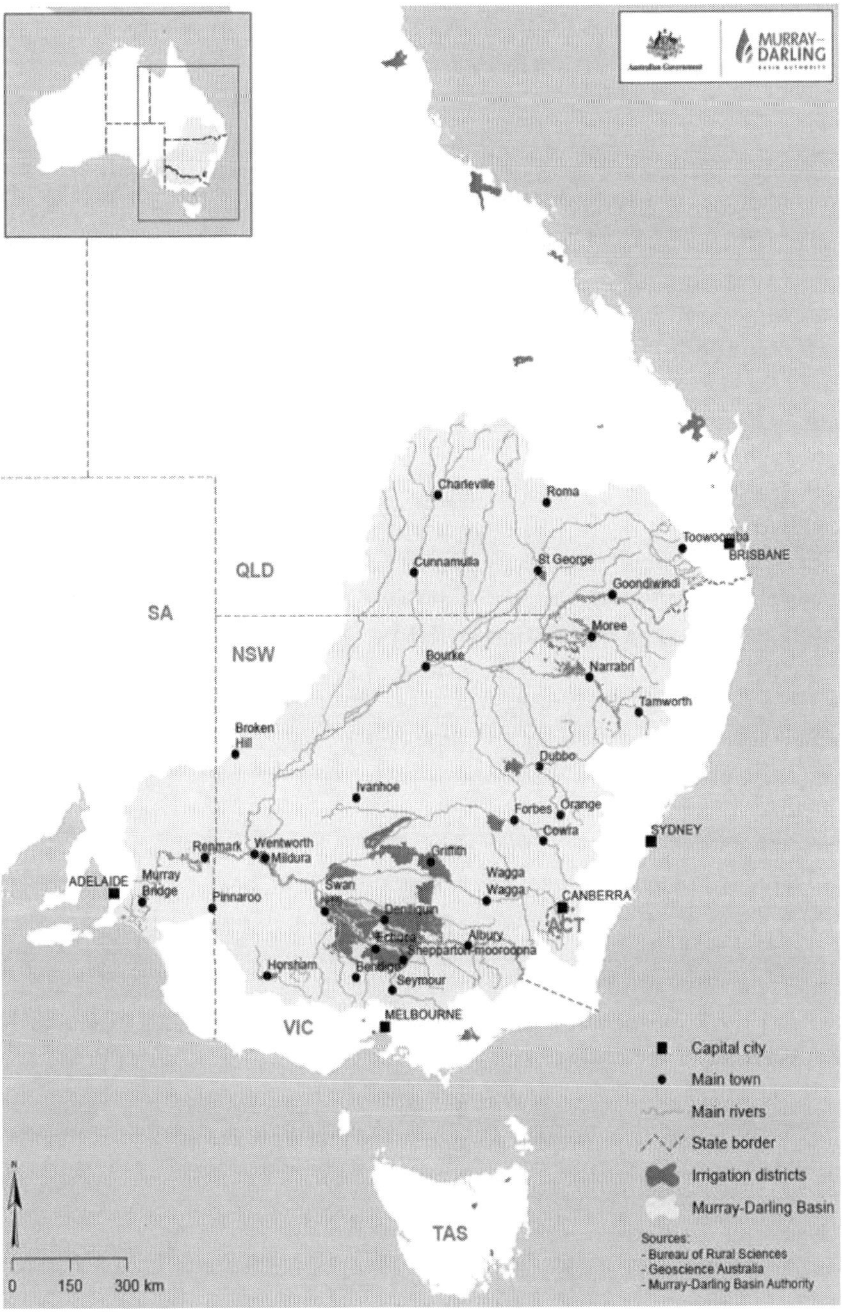

Figure 1. Irrigation in eastern Australia (Source: MDBA, 2010b).

3. Irrigation Water Demand within the Water Cycle

An understanding of the sources and sinks of water, including their interactions, is helpful in developing a suite of options for sustainability. Reviews are available on climate change and the global water cycle (Finlayson et al., 2011; Lawford, 2011). Several frameworks are available in Australia. The CSIRO Sustainable Yields Project (CSIRO, 2008) provides quantitative information on water balance at catchment scale including trends and projections for different emission scenarios and land-use change. The Regional Water Resource Assessment (NWC, 2007) provides annual national and regional water balances and provides estimates of certainty, noting that some items such as soil water content are more uncertain because they are derived by difference. River operators such as State Water in New South Wales provide catchment-based river system water balances that include assessment of 'unaccounted flow' such as evapotranspiration, surface water/groundwater exchange, unmeasured tributary inflows, river channel storage, and measurement inaccuracy and error (NSW DPI Office of Water, undated). Additionally, major reports on the resources of the Murray-Darling Basin are available (MDBA, 2010;Young & Chiew, 2011); and audits on surface and groundwater use and trading are provided by the Bureau of Meteorology (NWC, 2013b) for several key regions. The Cooperative Research Centre for Irrigation Futures provided a framework for policy and management interception for sustainable irrigation practice (Chartres & Williams, 2006).

Public participation is a key requirement for water planning (NWC, 2011a). It is supported by comprehensive information provided, in particular, by the National Water Commission, the Murray-Darling Basin Authority, CSIRO, the Bureau of Meteorology, several Cooperative Research Centres, State and Territory agencies, Universities and professional associations such as the Australian Water Association.

A holistic approach is required to capture the full range of options for irrigation sustainability that meet the criteria of public acceptability and cost effectiveness. An increasing concern for the ongoing management of water has been the deterioration of environmental assets because of excessive extraction of water for irrigation and other purposes. Some factors that need to be considered are listed below:

Firstly, there is great uncertainty over the meaning of environmental water. For example, it is assumed in the framework for water reporting recommended by the Independent Pricing and Regulatory Tribunal in New South Wales (SKM, 2005) that water 'left in the river' is water that is allocated to the environment even though the seasonality, duration and temperature of flow may be quite unsuitable for environmental benefit. Also water for irrigation is assumed to be separate from environmental water yet the same volume of water can deliver environmental, utilitarian and other benefits as it moves through the catchment (Hamstead, 2007; Syme & Nancarrow, 2008).

Secondly, water resource audits are generally annual summaries, yet variability from year to year is critical to selection of optimal irrigation cropping systems, to develop a hierarchy of access to water in water sharing plans and to optimising irrigation needs through 'carryover' and 'borrow' provisions for accessing water in storages between water accounting years. Water stored in the soil beyond a 'water year' is also important in assessing the efficiency and productivity of water use in irrigation cropping systems.

Thirdly, information on consumptive use of water needs to be updated to take into account the increasing demands of water for electricity generation, mining including coal seam gas extraction (NWC, 2010a; Williams et al., 2013), and the transfer of water from inland river systems to major urban centres and coastal cities. During the millennium drought (2002 to 2009) South Australia purchased water allocations to boost supply for Adelaide, and Victoria bought water for the rural cities of Ballarat and Bendigo and invested in a pipeline to deliver water from the MDB to Melbourne (Spies & Dandy, 2012). Consequently, in this chapter, options for water savings and reuse in urban and industrial situations, as well as in irrigation systems, are considered. Examples of innovation are provided in the following sections.

4. Maintaining Water Supply through Optimising Landscape Management

In Australia water supply is threatened by climate change and landscape changes, especially afforestation which has been extended to manage salinity and create carbon credits, but with the unintended consequence that run-off may be reduced in some catchments.

In the first national baseline assessment of unaccounted water use across the nation, the National Water Commission estimated the total unaccounted water use as a result of interception activities at about 5600 GL/yr consisting of about 2000GL/yr from forestry plantations, 1600 GL/yr from farm dams, 1100GL/yr from stock and domestic activities and 900GL/yr from overland flows used in floodplain harvesting (NWC, 2010b). In a wet year, interception volumes are estimated to be even greater; for example floodplain harvesting into farm dams from overland flows may potentially use up to 2600 GL. The total volume of unaccounted water is equivalent to roughly one quarter of all entitled water on issue (about 25 000 GL/yr). The Commission is establishing regulations to deal with unaccounted water use, and licenses for new forestry ventures are proposed.

Another approach is the optimised use of tree planting in priority areas to provide maximum amelioration of salinity with acceptable reduction of run-off flows. For example, the Victorian government commissioned consultants Sinclair Knight Merz to pilot this approach in the Loddon Catchment then extend it to several other catchments in Victoria (Meyer & Bowmer, 2004, page 114). Bowmer (2011) reviewed various models to optimise reductions in salinity while maintaining water run-off. Further information is provided by Nordblom et al. (2010); Nordblom et al. (2011); Vertessy et al. (2003); and Stirzaker et al. (2000).

5. Environmental Benefit with Minimal Use of More Water

5.1. Water Saving by Investment in Infrastructure

Estimates of irrigation water losses from irrigation supply and drainage channels by evaporation and seepage were estimated at 1500 GL in 2004-5, or about the same volume as that consumed by all Australia's major cities (ACIL Tasman, 2008). Losses can be reduced by installing new infrastructure to better measure and control flows to reduce evaporation and seepage and to replace outdated and leaky infrastructure. These savings can then help provide water for the environment at a lower social cost than government purchase of water entitlements. However the financial cost is high so this is a contentious issue for economists (Crase et al., 2013; Wittwer & Dixon, 2013). A large investment of AUD 5.8 billion has been committed to improving the efficiency and productivity of water use and management through irrigation modernisation plans (Australian Government Department of the Environment, undated).

The combined replacement value of the irrigation asset, consisting of control structures, channels and storage facilities, is estimated at more than AUD 15 billion, so managers are keen to make the most of these assets and almost all the major irrigation water delivery companies and private diverters have upgraded their supply systems in the past few years. Improved automation of surface irrigation has allowed a much greater control of water onto bays. Controls include pneumatic and electronic sensors positioned in bays close to the supply gate; timers that open and close irrigation gates at set times; and fully automated centralised control of multiple gates and channels using hydraulic control. Surge irrigation is a recent variation of surface irrigation that improves efficiency of row crop irrigation on freely draining soils. The major MDB companies, Coleambally Irrigation Cooperative Limited, Goulburn-Murray Water and Murray Irrigation Limited, have invested heavily in identifying excessively leaky supply channels and installing automatic control structures with remote monitoring and control (Meyer & Bowmer, 2004).

Ways of measuring water delivery at the farm gate are also changing. The robust Dethridge Wheel, while reasonably accurate when meticulously installed and maintained, is notorious for underestimating water volume in less than ideal conditions and has a delivery error of plus or minus 20 percent. It has been replaced in the Coleambally Irrigation Area by Rubicon's FlumeGate™, a device used to control and measure flow through a structure (flume) using differential pressure sensors. The Flumegate™ is automated, solar powered, safer to use, more accurate, cheaper to maintain and continuously monitored and controlled (see Total Channel Control below). In the Murrumbidgee Irrigation Areas Dethridge wheels are being replaced with pipe outlets equipped with ultrasonic depth meters. Murray Irrigation Limited is trialling other pipe meter solutions.

Throughout Australia, channel water supplies are being replaced and supplemented with pressurised pipe systems. Good examples are the rehabilitation of the Riverland irrigation areas in South Australia, the Angas Bremer scheme also in South Australia, and the Euston Cooperative high-pressure pipeline in south-western New South Wales.

The Australian, New South Wales, South Australian and Victorian governments contributed funds to an independent organisation, Water for Rivers, to recover 212 GL of water for the Snowy River and 70 GL for the River Murray. Water for Rivers used Total Channel Control (TCC) irrigation channel automation in the Goulburn-Murray Water irrigation scheme of north-central Victoria in pilot trials which were implemented by Rubicon

Systems Australia (Oakes, undated), though not without technical difficulties and problems with consultation and adoption (Race et al., 2007). Automation was applied to the main and secondary channel regulation gates and outfall structures and to outlets to individual farms where the manual Dethridge wheels were replaced by automated gates connected by a communications network, primarily based on radio technology. The farmers lodge and modify water orders directly with the TCC software system via the internet or telephone and the software accepts these orders in real time and operates the entire system to route water from the source to the point of demand. Later, Coleambally Irrigation recognised the opportunity to implement the solution in their entire 600 km channel system and the in-channel component was completed in 2005/2006 and delivered a system distribution efficiency of 90%. In subsequent years the farmer's outlets have been upgraded.

In 2005/2006 the Central Goulburn project demonstrated that in-channel automation could be combined with manually operated and controlled outlets, leading to automation of 3000 km of channel with more than 10,000 measuring devices connected to the system. The Shepparton Irrigation District documented a distribution efficiency of 90% of the 2010/2011 irrigation season. This approach has also been successful in the USA and piloted in China, hence demonstrating significant international opportunities.

Environmental benefits without the use of more water can also be achieved by reducing cold water pollution through modification of storage outlets, increasing longitudinal connectivity of rivers using fish ladders, opening weirs where possible, construction of weirs to prevent high summer flows in anabranches, restoring riparian vegetation and controlling alien fish (Lukasiewicz et al., 2013; State Water, 2013). McCartney (2009) provides an international review of the measures available to ameliorate the negative environmental impact of dams.

As noted by Cullen (2011), a critical issue for irrigation is the timing of investment in infrastructure and purchase of water for trading (see section 7 later). It is inappropriate to invest in refurbishing old irrigation districts where properties may be too small and irrigation layouts inappropriate for current irrigation technologies when a better strategy is to allow water to trade out and develop new enterprises on greenfield sites.

5.2. Water Savings by Optimising River Operations

Much water policy is concerned with 'cutting the cake' – water sharing between different users. However, optimisation of river operations with decision support systems such as computer-aided river management (CARM) can provide environmental benefit with minimal use of water. Other examples include flow pulsing, 'piggybacking' and countercyclical trading.

Computer Aided River Management is an approach to river operations being piloted in the Murray and Murrumbidgee Rivers that is supported by the Danish Hydrological Institute's suite of MIKE simulation models for key catchment and river processes. These include tributary inflows, river flow travel times, evaporation from open water, and evapotranspiration from riparian vegetation. In a proof of concept based on historical measurements, the CARM optimised solution was shown to significantly reduce dam releases without compromising irrigation water security. Based on information about the present state of the river, as well as optimisations conducted several times a day, river operators can run a 'leaner' river in which only water that is needed to meet demands is released. This reduces flow transmission losses and also leaves more room ('air space') for water storage in the river channel and in the ponds upstream of the weirs. This additional air space allows greater flexibility in operating the river weirs and provides storage for unexpected inflows and water orders that have been rejected by irrigators (van Kalken, 2013). The project recently received the Australian Water Association's National Award for Infrastructure and Innovation.

Water delivery efficiency is a key part of dealing with drought. Pulsing of flows by management of water releases from storages and weir pools has been used in northern New South Wales to reduce conveyance losses, but because of more variable river geomorphology, the technique is believed to be less useful in southern New South Wales and elsewhere. Pulsing of flow restores a degree of natural variability in hydrology that is important for resilient aquatic ecology, especially for the breeding of native fish (Watts et al., 2009). Also, by using small volumes of environmental water, connection of river and wetlands/floodplain can be achieved by 'piggybacking', adding an additional volume onto scheduled irrigation releases or unregulated tributary flow.

The Cooperative Research Centre for Irrigation Futures coined the phrase 'river harmonisation' to suggest an optimal solution for environment and consumptive use (Bristow & Stubbs, 2010). The approach includes changes in patterns of irrigation demand through conversion from summer to winter crops with the intention of reducing the damaging environmental effects of high a seasonal summer flow. Of course, a more comprehensive harmonisation would need to include management of all water sources (headwater storages, on-farm storages, unregulated supplementary water, rainwater, overland flow and groundwater) and optimise all the demands (consumptive, environmental and cross-border agreements). Statutory catchment-based water sharing plans and the Murray- Darling Basin Plan (MDBA, 2012) provide some of the required legislation and principles needed to achieve this challenge.

6. Water-Sharing Agreements

Water Sharing Plans are a critical component of water allocation for the environment and for meeting the social and economic needs of other users, particularly irrigators. They can help reduce conflict over water resources in stressed river systems and in climatic extremes. The National Water Commission recently published a report card for 157 individual water plans in Australia. Criteria for assessment include: planning for sustainable water extraction and overuse status; facilitation of water trade; integration of mining, forestry and other intercepting activities within the entitlement system; consideration of surface and groundwater connectivity; inclusion of accountable water management arrangements; the adequacy of monitoring, compliance and enforcement provisions; and evidence of planning for climate change (NWC, 2012).

States differ in the way they have approached water planning. For example, in New South Wales, Water Sharing Plans have been developed for regulated rivers and unregulated streams and rivers. Victoria uses a risk-based approach with sustainable diversion limits that are set regionally. Queensland uses Water Allocation Management Plans.

The emergence of 56 Catchment Management Authorities and Natural Resource Management groups across Australia provide a range of local approaches to planning. Of particular note to irrigators is the agreement on the hierarchy of rights to access water including the explicit 'right of the environment'. More recently 'the environment' has emerged as a customer with licensed allocation equivalent to consumptive users. Water for the environment has been obtained by Water for Rivers by improvements to river operating systems as described in section 5.2. Other schemes have been managed by Non Government Organisations including the Murray Wetlands Working Group, the Nature Conservation Water Trust and the World Wide Fund for Nature. In March 2008 the Commonwealth Government announced the first investment of AUD 50 million in the purchase of irrigation water for use as environmental flow.

The Murrumbidgee is an example of a river where excessive extraction has degraded the river and riverine ecology, making obvious the conflicting needs for irrigation and environment. Water sharing is particularly contentious because the river is highly regulated to provide water for irrigation that underpins regional wealth and employment. Water is also required to maintain the ecological function of the river and its floodplains, to support a Ramsar convention wetland and endangered fisheries, and to provide water for catchments downstream. The impact of irrigation extraction is felt both downstream and upstream. While the river is starved of water downstream there is over-supply to floodplains and wetlands upstream, as the river flows bank-full in summer from dam releases to supply irrigation several hundred kilometres downstream, reversing the pattern of natural flow.

The Murrumbidgee Water Sharing Plan establishes a hierarchy of rights for access to shares of the available water resource. Environmental water has priority, followed by landholders with riparian rights, then stock, domestic and utility needs, then irrigation water for permanent plantings (vines and fruit trees) and finally annual cropping requirements. Native (Aboriginal) title claims, still to be registered, are part of basic rights. Under the plan, water supply for towns (local water utilities) can be varied every five years to reflect changes in population and associated commercial activities, but water for new industry must be purchased through the entitlement and licensing system. The actual volume or share of consumptive water depends on the announced allocation of water, storage volumes and climatic conditions. Consequently, any reduction in supply, for example as a result of reduction of catchment run-off through afforestation or climate change, is reflected in a reduction of share to the lowest priority of access, the supply to annual cropping.

Environmental water is managed through a series of flow rules which include a minimum low flow from the two major storage dams. 'Translucency' rules affecting dam outflow releases provide a proportion of inflowing water to reflect some of the natural variability in the release pattern. The aims are to minimise the accumulation of sediments, control algal blooms caused by low flow, and restore some of the early spring flows that are important for fish breeding. An 'environmental health allowance' is a volume of water set aside in storage for opportunistic release for environmental benefit that includes water for 'end-of-system' flow to ensure connection with the River Murray downstream. Novel aspects of the flow rules include 'piggy backing' and 'provisional storage' rules. In 'piggybacking' arrangements the environmental health water allowance is used to raise river height to inundate wetlands, especially in the middle reaches of the river, by supplementing dam releases onto surge peaks from tributary inflows. In addition, a volume of water ('provisional water') is retained in the dam and increases as the water allocation increases. This increases the probability of a dam spill in winter and spring, and also the level of storage at the start of the following year. This effectively shares environmental water with irrigators in wetter times to reduce the highest year impact on irrigators and improves early season allocations (NSW Government, 2003).

7. Water Markets and Adaptation to Water Scarcity

Over-allocation of water resources to irrigation, industry and cities has severely damaged flow-dependent riverine ecosystems and led to growing interest in ways to restore water to the environment beyond the water sharing plans described above. Water buyback has become increasingly popular in Australia and the USA (Lane-Miller et al., 2013; Wheeler et al., 2013).

Australian rural and irrigation communities have faced a range of challenges in the past 40 years including a decade of major drought followed by several years of flooding, declining terms of trade, an aging workforce, population loss, business closures, and the withdrawal of education and health services from communities that no longer achieve 'critical mass' to justify government investment. Consequently many rural communities fear for their future (Wheeler et al., 2013). In this context the development of water markets are increasingly beneficial in assisting irrigators to adapt to climate change and other pressures.

Water markets are likely to increase economic efficiency because prices make the opportunity cost of water explicit (Qureshi et al., 2009). They also provide incentives to adopt water-saving technologies and reduce inefficient uses of water. The gain from water trading in a dry year was estimated at AUD 495 million (Peterson et al., 2004) while water trading in the southern MDB increased Australia's gross domestic product by AUD 220 million in 2008–2009 (NWC, 2010c).

Several methods are being trialled to balance the needs of irrigators and the environment through trading. A process, developed by the Murray Wetlands Working Group, provides benefit to both wetlands and irrigators.

Investment in seepage reduction was used to provide water on private wetlands, and water was purchased for the environment during wetter years when prices were lower (NSW MWWG, undated). In another proposal, countercyclical trading, water is provided to the environment above a given allocation in wetter years in exchange for increased irrigation security in drier years (Heaney et al., 2004; Kirby et al., 2006). The Commonwealth Environmental Water Holder has recently announced that some water previously purchased for the environment will be offered for sale to irrigators to help cope with the re-emergence of very dry conditions (ABC, 2014).

A recent overview of the status and performance of the water market in Australia is available from the National Water Commission. Some improvements in the quality and timely provision of information on prices and volumes traded would encourage water market participation and lower uncertainty, but progress is generally positive (NWC, 2013a).

8. Improved Water Use Efficiency on-Farm

8.1. The Meaning of Water Use Efficiency

There is a global focus on improved water use efficiency to deal with water scarcity. In Australia terms and definitions have been clarified (Purcell & Currey, 2003). Meyer et al. (2005) caution against the use of terms such as ML/ha or $/ML as efficiency measures, noting the importance of recognising efficiencies at storage, delivery and field application levels. The fate of water diverted from the Murray and Murrumbidgee Rivers illustrates the need for a holistic approach to assessing efficiency: of 8608 GL diverted, 78% was delivered to the farm gate, 60% was applied to crops and only 42% was used productively through crop transpiration (Meyer & Montagu, 2006). Water applied to the field is lost by non- beneficial evaporation while transpiration associated with plant dry matter, fruit and grain production is the output that is sought. This beneficial production per unit of water (water use productivity kg/ML) is proposed as a better estimate of irrigation system effectiveness (Meyer et al., 2005). In the whole MDB, losses close to 5000GL, approaching half the water diverted, are caused by seepage and deep drainage, evaporation from open water surfaces and from the soil, requiring a holistic approach to consideration of water use. Storage and system options have been described previously (section 5).

The following sections give some examples of technology that can save water. However, as noted by Humphreys et al. (2005), the concept of water saved can be elusive because some seepage, percolation and run-off losses can be recaptured, for example through groundwater pumping. Also it should be noted that annual crops such as rice provide opportunistic use of water in wetter years, providing the resilience and capacity to adjust to climate variability (Meyer et al., 2005).

8.2. Land and Water Plans

As noted by Meyer et al. (2005) the development of Land and Water Management Plans in every major region in the Murrumbidgee and Murray Basins illustrates the increasing connection between irrigators and their role in land and water resources.

For example, Murray Irrigation Limited (MIL) operates Australia's largest privately owned irrigation company, delivering 1.2 million ML of water to 2,400 landholdings in southern New South Wales each year. The company needed to manage rising watertable and the associated threat of salinity by encouraging water use efficiency and best management practices. MIL have introduced a world-leading total farm water balance policy and set limits on irrigation intensity to prevent long-term degradation of the groundwater and soil resources. SWAGMAN Destiny, a decision-support software program, was used to map the intensity of water application that could be made safely and avoid excessive groundwater accessions (Prathapar et al., 1997). The policy is part of the extensive Land and Water Management Plan for the region, which includes components to restore significant wetlands and remnant vegetation. In related activities since 1995, the company has also spent: AUD 3.45 million on seepage control and channel sealing to decrease irrigation delivery system losses, and around AUD 16 million on telemetry systems that monitor and provide control on channel and diversion structures. Combined with an education program, irrigators and other community members are taking renewed responsibility for their area. Watertables are generally falling, wetlands are receiving managed water supplies and there are many revegetation projects underway. In 2003 the company and community effort was recognised with an Australian Water Association Environmental Merit Award (Meyer & Bowmer, 2004).

8.3. Laser Leveling, Layout Design and Improved Irrigation Systems

Surface irrigation is still the main form of irrigation in Australia, although with water supplies becoming more valuable, there is increased interest in using more controlled forms of irrigation, especially on high-value horticultural crops. For some soil types and well-designed irrigation layouts, well-managed surface irrigation can minimise loss of water into the deeper soil layers that are beyond most plant roots. The advantages of improved surface irrigation design and layout are often fully realised when they are implemented within a whole-farm plan to optimise use of soils and design the best layout for supply and drainage systems. For example, in Queensland, DamEa$y is a decision support system that improves irrigation water management by optimising the combination of system water availability and on-farm storage (Narayan et al., 2007). In northern New South Wales consultants are prominent in the design of recycling systems in the summer rainfall cotton areas. Water is often harvested from overland and high river flows and stored in large dams and 'ring tanks'. Well-designed distribution channels, laser graded irrigation bays and recirculation storages collect drainage run-off and rainfall run-off to improve water use productivity (Raine et al., 2005).

The latest development in this area is the 'bankless channel' system (Grabham et al., undated). The bays are graded flat, with a rooftop in the centre of the bay and a gradient of 1:3500 from the centre to the bankless channel, with bays stepped down the slope. The layout is suitable both for crops grown on the flat, such as rice, and row crops such as maize. Benefits include less labour, less drainage, more even application of water across the field, deep ponding for weed control during a rice season and automation of water control.

Laser-assisted levelling combined with layout design has been a boon to irrigation, in particular to surface irrigation. Good grades improve the evenness of distribution and speed of irrigation, and reduce labour costs to manage water. Australian operators have optimised layout design to minimise the volumes of soil that need to be shifted, and to deliver water efficiently around the farm. Global positioning systems (GPS) are used on earth-moving plant to give very precise grades with a minimum of soil disturbance. Irrigation surveyors have greatly assisted in improving surveys of soil and topography and in providing design and construction plans for effective surface irrigation.

Fairweather et al. (undated) reviewed new techniques for improving the effectiveness of all types of irrigation systems including centre pivot and linear move systems, partial root zone drying, irrigation scheduling using remotely sensed crop temperatures, deficit irrigation and alternate furrow irrigation.

8.4. Precision Farming

Electromagnetic surveys (EMS) are used in whole farm planning to provide information about where salts are stored in the soil (Cresswell et al., 2007). EMS can also identify suitable land for drains, on-farm channels, storages, paddock irrigation design, cropping and pasture planting, as well as providing information on where to undertake remediation methods such as applying gypsum. The methods have been developed and tuned for viticulture (Lamb et al., 2004) and rice (Beecher & Dunn, 2005).

Precision farming uses GPS in combination with EMS and airborne video mapping or satellite imagery. Increasingly, producers, professional agronomists and water suppliers are able to map areas of different crops and understand variations in soil character and water within fields and design land and farming systems to match the variation (Lamb, 2000). This technology underpins a redesign of agriculture and support the concept 'double the yield in half the area' (Meyer, 2003) that has been proposed as a target. Many viticulturalists use precision farming commercially to improve water use efficiency and improve grape quality (Hall et al., 2002). Interest in Australian technology is being shown by vignerons in Chile, New Zealand, South Africa and the USA.

8.5. Soil Water Monitoring

Australian technology has led the world in the recent development and application of electrical capacitance sensors that are used in a similar way to neutron soil water probes. Their advantages are reduced cost, no radiation hazard, and the ability to continuously monitor changes in water content through the soil profile. Sentek Technologies, based in Adelaide, South Australia has pioneered the development of an extensive range of sensors, which now come with user-friendly software to assist in analysis of real-time information and in irrigation management (Jabro et al., 2005).

An alternative way of managing irrigation is to measure when the soil has sufficient water for plant growth. A deceptively simple device, FullStop™, enables the soil-wetting front to be measured through location of a pair of detectors, one halfway down and another near the bottom of the root zone. A visible signal at the surface helps farmers adjust irrigation to plant needs as they grow through the season (Stirzaker et al., 2010). Fullstop™ allows farmers to share information on best practice and develop their own accreditation system. The technology is owned by CSIRO Land and Water and is marketed in South Africa by Agriplas Ltd under license, with further markets being developed in South Africa, Europe, China and South America (Agriplas, 2009).

8.6. Irrigation Industry Developments in Water Use Productivity

In Australia, a variety of new irrigation methods and techniques have been developed by CSIRO, state agencies, Cooperative Research Centres, irrigation corporations, private irrigation agronomists and consultants, and by enterprising farmers themselves. Several commodity group industries have proudly demonstrated improvements in water use and crop quality over recent years.

Water use in sugar cane, rice, cotton, fruit grapevines, vegetables, and other broadacre crops is substantial (Table 1; ABS, 2010). Some examples of innovation in selected industries are described below.

Dairy

Dairy pasture is the single biggest user of irrigation water in Australia. Large dairy areas, such as the Goulburn region around Shepparton in Victoria, are highly dependent on assured water for irrigation. Murray Dairy, one of the regional groups associated with Dairy Australia, has worked closely with the Goulburn Broken Catchment Management Authority to improve irrigation practice and cope with the recent drought. On-farm works include converting unimproved flood irrigation practices to improved flood irrigation practices through laser grading and levelling, purchase and installation of irrigation scheduling equipment and conversion to other new technologies (Murray Dairy, 2012). The Victoria Department of Primary Industries at Tatura has compared irrigation methods and found that sprinkler and subsurface drip is more efficient than border check and surge irrigation methods (Wood &

Finger, 2006). The emphasis on best management practices means that the dairy industry is well placed to continue with innovation and adaptation.

Viticulture

Best practice irrigation management (Tee et al., 2003) has been described by the Cooperative Research Centre for Viticulture (now hosted by the Australian Wine Research Institute). Shortage of water is the factor most likely to limit the growth of the Australian wine industry. Regulated deficit irrigation (RDI) and partial root zone drying (PRD) (Dry et al. 1996), are techniques that save water while improving production and grape quality (Possingham, 2002).

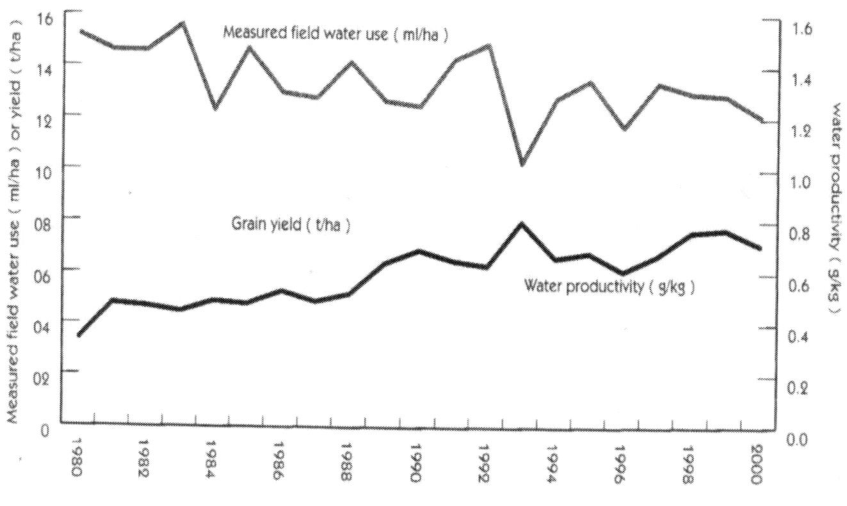

Figure 2. Rice field water use, grain and water productivity over time (Humphreys et al., 2003).

Rice

Typically, up to 150 000 hectares of temperate varieties of rice are sown each year in the Murray Valley of New South Wales and Victoria, and in the Murrumbidgee Valley of New South Wales, producing about1.2 million tonnes of paddy rice. Including flow-on effects, the industry generates over AUD 4 billion annually for regional communities and the Australian economy. Over the past twenty years, field irrigation water-use productivity has more than doubled (Figure 2) as a result from an average of around 0.34g paddy rice per kg water to around 0.77 g/kg, largely due to increased yield from the development and adoption of improved varieties and management strategies, and to a lesser degree due to the introduction of rice water use and soil suitability policies (Humphreys et al., 2006).

Water use has also been reduced by restricting rice growing to less leaky soils, by reducing the length of ponding of floodwaters, and by using precision agriculture to manage salinity and fertilisers. The industry has developed the *Environmental Champions Program* to accredit growers and encourage them to maintain and develop best practice (RGA, 2014). Progress in integrating science, people and policy makers was recognised in 2002 by the award of the first 'Reference Catchment' status to the lower Murrumbidgee catchment, a major Australian rice-growing region, under the UNESCO/WMO program HELP (Hydrology for Environment, Life and Policy) (Humphreys et al., 2006).

Cotton

By world standards, the Australian cotton industry ranks highly in water-use productivity, producing high yields per hectare and per ML of water (Figure 3). Roth et al. (2013) provide a recent review of progress achieved by the Cotton Catchment Communities Cooperative Research Centre Program. The industry has invested in a breeding program for high yield and pest resistance. Water management is tightly controlled through laser leveling of fields, water recycling, improved use of weather forecasting, and precision scheduling. Most cotton soils are suited to surface furrow irrigation, with drip and lateral move overhead irrigation systems used to an increasing extent. Farm and channel design, and the use of on-farm water storages and off-farm water infrastructure, are all carefully managed to optimise water use. Australian cotton growers are optimising water use and yield using irrigation scheduling and furrow irrigation automated application systems. Hydrologic software developed by CSIRO Plant Industry with the Cotton CRC integrates information on soil water, cotton boll numbers and leaf area and with long-term climate data to predict crop yields and advise on optimum dates for irrigation (Richards et al., 2008).

8.7. Improved Skills Training and Irrigation Education

The critical role but associated poor proficiency of skills in the irrigation industry was highlighted in the early 1990's. There was recognition that all aspects of irrigation were poorly served from system design, supply installations and operations, on-farm application and crop management.

A concerted skills training and accreditation process was started and has continued largely through the Irrigation Association of Australia. At about the same time, audits of irrigation education opportunities, particularly at tertiary level indicated that these were in need of significant updating and improvement (Meyer & Taylor, 1998). The combined effect of these efforts has seen significant improvement in the quality of service provision to irrigators in terms of design, installation and management of irrigation systems along with a much greater awareness of responsibilities to use water in a productive manner that minimises the effects on supplying and receiving ecosystems.

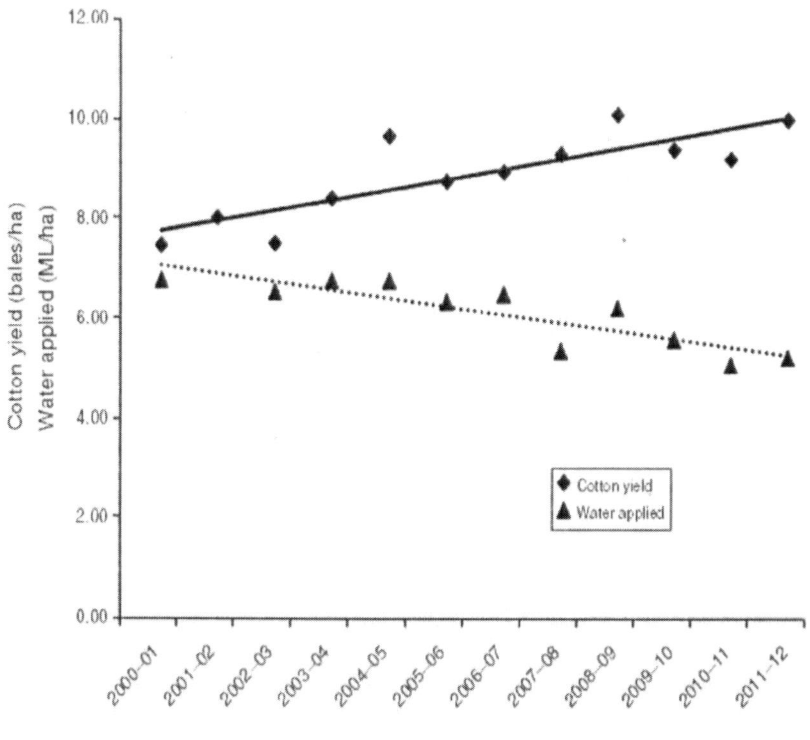

Figure 3. Irrigated cotton yields and irrigation water applied in Australia, 2001-2012 (Roth et al., 2013).

It is increasingly appreciated that irrigation sustainability will occur within the nexus of water, costs and knowledge. Improved knowledge is associated with better and more informed management. It is also highly dependent on the continued understanding of the basic processes associated with soils, water and plants and on continued innovation of irrigation practice. Research and education is therefore fundamental to improved irrigation. Two significant National programs contributed to this need. The National Program for Sustainable Irrigation (http://www.npsi.gov.au/) and the Cooperative Research Centre for Irrigation Futures (http://www.irrigationfutures.org.au/) brought together government agencies, research and development agencies, education and training providers and irrigation water supply companies to focus on new developments to improve irrigation. From 2002 to 2009 the Cooperative Research Centre for Irrigation Futures (CRC IF) invested AUD 72 million from 14 research, education and industry partners. CRC IF was formed to facilitate cooperative research networks and programs that continuously improve irrigation policy, tools, practices and processes to double irrigation water use efficiency, improve profitability for commercial irrigation enterprises and protect and enhance landscapes and the environment. A significant success of CRC IF was the training of 50 post graduate students who have become widely distributed within the water and irrigation sectors.

8.8. Combined Use of Surface and Groundwater

Greater competition for water, compounded by drought during the decade to 2010, has focused attention on increasing water-use productivity and on using groundwater in conjunction with surface water. Groundwater replenishment (recharge) rates are often low, so careful management is required to avoid over-exploiting the supply. Several examples of changed irrigation practices and substitution of groundwater with alternative water sources are available.

In the Lower Burdekin, Queensland, a range of partners are working together to manage surface water recharge and groundwater extraction to prevent the encroachment of seawater into the groundwater supply. In the Virginia District, about 35 km north of Adelaide the market garden area is expected to increase by 2 000 hectares, using 30 GL of reclaimed urban effluent from the Bolivar Virginia pipeline and 18 GL of water from the underlying aquifer. This is a good example of the use of reclaimed water to replace the use of overexploited groundwater. In the McLaren Vale region south of Adelaide a group of vignerons formed the Willunga Basin Water Company to build, administer and operate a pipeline to access reclaimed water from the Christies Beach Wastewater Treatment Plant. The reclaimed water comes from an urban area that services over 150 000 people and treats 27 ML of household and industrial effluent each day. The Class B reclaimed water is suitable for drip and subsurface drip irrigation, but cannot be used for spray irrigation. The Willunga Basin Water Company has implemented safety measures to minimise the health risks of using reclaimed water. The company provides 3 GL of reclaimed water to the growers each year. The success of this venture has ensured that other regions, particularly those with ready access to large urban areas, will consider using reclaimed water as a valuable source of irrigation water (Meyer & Bowmer, 2004).

8.9. Re-Use of Drainage Water and Effluents for Irrigation

As the demand for water grows, novel approaches are being employed to harness and reuse water. The subject has been reviewed for the Australian Academy of Technological Sciences and Engineering by Radcliffe (2003).

Treated effluent from urban and industrial areas provides a small but increasing component of the water used for irrigation. Melbourne releases 12 GL of treated effluent to inland rivers every year, and 327 GL to coastal waters. Sydney releases 44 and 368 GL respectively. Re-use has the dual benefits of environmental protection and resource recovery. Treated wastewater from large cities is low in pollutants and pathogens and contains nutrients that benefit agricultural production. It can be used successfully if irrigation systems are designed to minimise direct water contact with plants, and salt loads are carefully managed. New enterprises in Werribee near Melbourne and Bolivar near Adelaide successfully use recycled water from urban areas for intensively managed horticultural crops. Consultants and engineers specialising in appropriate technology are accredited by the Irrigation Association of Australia. Managing public and industry perceptions underpins acceptable water reuse.

Irrigation water re-use also provides protection for the environment. In Australia the quality of irrigation drainage waters (pesticides, salinity, turbidity, nutrients and pathogens) is strictly monitored by environmental protection agencies. For some rice farms, licences require that the first rainfall run-off be captured and recycled. For cotton, the toxicity of the pesticides used drove regulators to require a closed system with on-farm recycling of tailwater. While water quality concerns initially drove the move to recycling, farmers now see advantages in water saving, and recycling has become standard practice.

Effluent waters from intensive rural industries, such as poultry or piggery operations, abattoirs, wineries and urban sewage are also being recycled in irrigation with the twin objectives of avoiding pollution and turning waste into wealth. Manuals and interactive decision-support software package for spreading a nutrient-rich waste on agricultural land are available. These are suitable for piggeries, cattle feedlots, dairy sheds, sheep feedlots, winery and distillery wastes, poultry shed wastes and processing of olives, fruit and vegetables (Bond & Smith, 2006).

FILTER (Filtration and Irrigated Cropping for Land Treatment and Effluent Reuse) is a novel low-tech solution that combines the use of nutrient-rich wastewater for intensive cropping with filtration through the soil to a subsurface drainage system (Jayawardane et al., 2001). It can be used in tandem with Sequential Biological Concentration, which in addition, concentrates the salt into a manageable volume for use in aquaculture or evaporation. FILTER has been demonstrated in partnership with Griffith City Council, New South Wales, for treating secondary sewage; with Riverina Wines for winery effluent; and with domestic effluent schemes in India and China.

9. More Efficient Water Use and Re-Use in Industry and Cities

Competition for water from cities and industry is emerging as a threat to sustainable irrigation. As noted by the Australian Academy of Technological Sciences and Engineering (Spies & Dandy, 2012) transferring water from rural to urban areas by trading and pipelines is logical given that a relatively small fraction of agricultural use represents a large fraction of urban demand. In Victoria a network of rivers, storages and pipes have been created to manage water as an integrated resource. A 70 km River Murray to Melbourne pipeline was built recently at a cost of AUD 750 million to provide 75GL/yr of water to the Melbourne city network. With improved rainfall in the Melbourne catchments the pipeline supply has not been used, allaying concerns that irrigation water would be transferred to urban users, even though water savings would be derived from the Northern Victoria Irrigation Renewal Project (Spies & Dandy, 2012). A very long pipeline from the Ord Irrigation Scheme in northern Western Australia to Perth has also been proposed but costs are extremely high (Marsden Jacob Associates, 2008). With an open water trading market the number of rural to urban trades is likely to increase as demand and connectivity increases (NWC, 2011b).

As well as water trading and use of pipelines many alternatives are being developed. These include, in order of increasing cost, 'catchment vegetation thinning' to increase run-off, demand management, re-use of stormwater, increased use of local groundwater, large scale non-potable re-use including water scheduling for urban gardens, local aquifer recharge and re-use, desalination of seawater, use of water efficient domestic appliances, reduction of conveyance losses by replacement of pipes and infrastructure, and use of rainwater tanks (Marsden Jacob Associates, 2008; Gregory & Hall, 2011). Many of the approaches are integrated in a new paradigm of planning called 'Water Sensitive Urban Design' (BMT WBM Pty Ltd 2009; Wong & Brown, 2011).

The direct reuse of treated effluent water for drinking is controversial (Lampard et al., 2010) but is advocated by the National Water Commission (GHD, 2007). As noted by Troy (2011) the emphasis on finding new sources of water for cities must change because all the options are expensive and some damage the environment. He advocates greater attention to the supply of potable water and ways in which wastes are managed, noting that water used for maintenance of sewage systems accounts for almost half that used inside the house.

Use of water in mining and fracking to extract coal seam gas are recent threats to groundwater aquifers and to pollution of surface water (Williams et al., 2013). These issues are not covered in current water sharing plans. Land use planning is proposed to protect the most valuable agricultural land and most vulnerable water systems. This is currently a highly contentious issue in Australia.

10. Public Participation

Developing priorities for investment in water resource protection requires that biophysical considerations should be better integrated with socio-economics. Notably, the effect of public expectation on driving or restricting the implementation of policy must be considered. These issues are complex, difficult and uncertain (Ross et al., 2012; Syme, 2013). Advantages of community involvement are that people can enrich expert knowledge, set priorties for limited volumes of water, find local innovative solutions and agree on monitoring regimes to demonstrate environmental benefit.

In discussing the contentious issue of water allocation, Syme (2013) reports that people generally support concepts of long-term sustainability and inter-generational equity but uncertainty about the future requires that trust and fairness are explicitly included. Trust often reflects the legacy of ongoing interaction between government departments and community and fairness is an overall concept involving both procedural and distributive justice.

As noted earlier, critical issues in Australia include the effect of a ten year drought on shortage of urban and industrial water, and development of a plan for the Murray-Darling Basin to increase the water available for the environment. Release of the proposed MDB Plan met with a strong and hostile reaction (Quiggan, 2012). Much of this response came from the fear that decreased water for irrigation would severely reduce local and regional economic and social viability. While some of this concern is valid much of it resulted from under- estimating the innovation, adaptation and water productivity improvement that is possible and previously demonstrated. To their credit, planners and governments are responding to these challenges.

Fortunately, a wide range of tools is being developed and sharing of knowledge is enhancing collaborative approaches to water planning. These include planning guidelines (IACSEA, 1998); a systems analysis template for integrating across social and environmental disciplines (Newell et al., 2005), multi-criteria analysis (Hajkowitz, 2007) and multi-objective optimisation methods (Xevi & Khan, 2003). The Journal of Hydrology (Anon, 2012) recently published a special issue of case studies and Tan et al. (2010) provide aids to select practical guides and tools for water planning that include stakeholder analysis, indigenous engagement, socio-economic impact assessment, best practice for managing climate risk, participatory mapping, and deliberative multi-criteria evaluation.

Also, in the last decade strategies and programs for watershed protection have evolved in Australia into strategies and programs for Integrated Catchment Management, Total Catchment Management and Whole Catchment Planning (Ryan et al., 2010). These approaches provide a framework for public participation in decision making and priority setting. For example 'resilience thinking' has been applied to the several catchments in New South Wales and Victoria to engage communities in planning for natural variation and extreme climate (Walker & Salt, 2012). As noted by Ross et al.(2012), in the face of change, a resilience perspective is required to foster each community's and region's strengths so they can plan effectively and make their own adaptations. An important development in effective community engagement and future planning is the increasing recognition that identifying shared values is fundamental to successfully implementing adaptive changes (Hatton MacDonald et al. 2013). The fundamental importance of identifying stakeholder values as the major influence in likely success of change in complex social ecological systems is explored by Wells and MacLean (2013). For operating in 'the paradigm of complexity', they set out a methodology ('One Way Forward') that has four central components – 'envisioning, core messages (values), indicators of progress, and experimentation'. Developing and applying this understanding of social change processes should greatly assist irrigation area communities to continually adapt to more sustainable systems.

11. Assessment of Progress towards Irrigation Sustainability

Over the ten years of the Millennium drought in Australia there was a 70% reduction in irrigation water use but the gross value of irrigated production fell only by 14%, mainly reflecting increased prices for dairy and cereals but also because of various adaptation strategies employed by irrigators (Kirby, 2011).

An independent review of Australian water management by the Organisation for Economic Cooperation and Development (OECD, 2008) recognised substantial progress in the implementation of water reforms with the adoption of the National Water Initiative in 2004. These reforms include the separation of land property rights and water access entitlements in all states and territories and the development of institutional arrangements for water trading. The new arrangements deal with the environmental constraints imposed by the predominantly dry and variable climate in Australia by setting environmental flow regimes and by defining water rights as a share of the consumptive pool. OECD commended the development of water access entitlements, institutional arrangements for trading and setting environmental flow, development of an integrated catchment-based approach, and separation of regulatory and water service delivery functions.

In water reform the restructure of communities affected by a reduction in water supply remains contentious. This chapter has described substantial progress being made in optimising water availability for all consumers, options for achieving environmental benefit without the use of more water and improved water use productivity through technical developments in irrigation agriculture and river operations.

Australians have an expectation that they will be involved in decisions that affect their welfare and livelihood. Much thought has been given by social scientists and economists on how to achieve public participation in environmental planning (Lane & McDonald, 2005) and how to redesign Australian agriculture (McKenzie & Williams, 2013, Bryan et al., 2013). Though transaction costs of community consultation are high, and the 'rush to increase public subsidy in infrastructure to appease those claiming to be disaffected' has been criticised (Crase et al., 2013) the evidence presented in this chapter supports the premise that re-allocation of water is being achieved while irrigation remains viable and sustainable.

CONCLUSION

Australian irrigation expertise covers a range of industries and climates, from tropical to temperate. Long-established gravity-based systems are changing to new pressurised high-technology irrigation systems with specialised precision farming. Experienced consultants, engineers and planners are developing environmentally-sensitive solutions through improved design and management of supply and drainage systems. Agronomists and scientists are developing new cropping systems that are more productive in using water while retaining profitability and regional employment.

Irrigation in Australia is embedded in a holistic policy and planning framework that considers the whole water cycle and includes a commitment to integrated catchment management and equitable water sharing principles. Australia is already acknowledged as a world leader in community participation in integrated catchment management through the Landcare movement, and the development of Land and Water Management Plans and Water Sharing Plans at community level. An integrated approach is required to evaluate options, prioritise investments, maximise economic returns, guide policy and balance the environmental demands of river ecosystems with the needs of irrigated agriculture and its dependent regional communities. This chapter has described the progress being made through the implementation of national water policy and the development and application of innovative technologies at farm, irrigation area and regional scale. The competition for water is fierce, but innovative technologies and planning approaches in both irrigation and urban water use, together with the introduction of water trading are providing substantial national benefits.

Acknowledgment

Information provided in sections 5.1, 8.3-8.6 and 8.8-8.9 is adapted from a previous publication (Meyer & Bowmer, 2004).

References

ABC. (2014). Murray-Darling water allocations to be sold back to farmers after years of environmental buy-backs. Retrieved 20 January 2014, from http://www.abc.net.au/news/ 2014-01-20/murray-darling-water-licences-to-be-sold-back-to-farmers/5207632

ABS (Australian Bureau of Statistics). (2010). Water Use on Australian Farms 2008-9. Retrieved 17 January 2014, from http://www.abs.gov.au/ AUSSTATS/abs@.nsf/DetailsPage/4618.02008-09?OpenDocument #Publications

ACIL Tasman. (2008). Australia's Working Rivers. The Role of Infrastructure and Water Buybacks in Recovering Environmental Flows. Melbourne, Australia. Retrieved 17 January 2014, from http://www.infrastructureaustralia.gov.au/public_submissions/ published/files/445_cranegrouplimited_SUB2.pdf

Agriplas. (2009). Agriplas Wetting Front Detector Retrieved 16 January 2014, from http://www.agriplas.co.za/index.php/product/wetting-front-detector/

Anon. (2012). *Journal of Hydrology, Special Issue Series, 474*.

Australian Government Department of the Environment. (Undated). Irrigation Modernisation Retrieved 17 January 2014, from http://www.environment.gov.au/topics/water/rural-water/sustainable-rural-water-use-and-infrastructure/irrigation-modernisation

Beecher, G. & Dunn, B. (2005). EM31 or EM38 for rice soil assessment. *IREC Farmers Newsletter, 170*, Spring 2005, 56-57. Griffith , NSW, Australia:. Irrigation Research and Extension Committee

BMT WBM Pty Ltd. (2009). Evaluating Options for Water Sensitive Urban Design- A National Guide: Joint Steering Committee for Water Sensitive Cities. Retrieved 17 January 2014, from http://www.environment.gov.au/system/files/resources/1873905a-f5b7-4e3c-8f450259a32a94b1/files/ wsud-guidelines.pdf

Bond, W. & Smith, C. (2006). Environmental implications of reclaimed water use in irrigated agriculture. In D. Stevens (Ed.), *Growing Crops with Reclaimed Wastewater* (171-179). Collingwood, Vic, Australia.

Bowmer, K. H. (2011). Water resource protection in Australia: Links between land use and river health with a focus on stubble farming systems. *Journal of Hydrology*, *403*(1-2), 176-185.

Bowmer, K. H. (2013). Ecosystem effects from nutrients and pesticide pollutants: Catchment Care as a solution. *Resources*, *2*, 439-456. doi: 10.3390/resources2030439

Bristow, K. L. & Stubbs, T. (2010). *Re-inventing Irrigation Catchments. The System Harmonisation Story*: CRC for Irrigation Futures. Retrieved 17 January 2014, from http://www.piarn.org.au/sites/piarn.boab.info/files/KBristow%20reading%20-System%20Harmonisation%20Book.pdf

Bryan, B. A., Meyer, W. S., Campbell, C. A., Harris, G. P., Lefroy, T., Lyle, G., Martin, P., McLean, J., Montagu, K., Rickards, L. A., Summers, D. M., Thackway, R., Wells, S. & Young, M. (2013). The second industrial transformation of Australian landscapes. *Current Opinion in Environmental Sustainability*, *5*, 278-287.

CEWH (Commonwealth Environmental Water Office). (2013). 2011-12 Outcomes Report. Canberra, Australia. Retrieved 17 January 2014, from http://www.environment.gov.au/ resource/commonwealth-environmental-water-office-2011-12-outcomes-report

Chartres, C. & Varma, S. (2010). *Out of Water. From Abundance to Scarcity and How to Solve the World's Water Problems*: Upper Saddle River , New Jersey, USA: Pearson Education Ltd.

Chartres, C. & Williams, J. (2006). Can Australia overcome its water scarcity problems? *Journal of Developments in Sustainable Agriculture*, *1*, 17-24.

COAG (Council of Australian Governments). (2013). *Intergovernmental Agreement on Implementing Water Reform in the Murray-Darling Basin.* Retrieved 17 January 2014, from https://www.coag.gov.au/sites /default/files/Intergovernmental%20Agreement%20on% 20Implementing%20Water%20Reform%20in%20the%20Murray%20Darling%20Basin%20%28signed%29.pdf.

Crase, L., O'Keefe, S. & Dollery, B. (2013). Talk is cheap, or is it? The cost of consulting about uncertain reallocation of water in the Murray–Darling Basin, Australia. *Ecological Economics*, *88*, 206-213. doi: http://dx.doi.org/10.1016/j.ecolecon.2012.12.015

Cresswell, R. G., Mullen, I. C., Kingham, R., Kellett, J., Dent, D. L. & Jones, G. L. (2007). Airborne electromagnetics supporting salinity and natural resource management decisions at the field scale in Australia. *International Journal of Applied Earth Observation and Geoinformation*, *9*(2), 91-102.

Cribb, J. (2010). *The Coming Famine: Risks and Solutions for Global Food Security*. Berkeley, California, USA: University of California Press.

CSIRO (Commonwealth Scientific and Industrial Research Organisation). (2008). Water Availability in the Murray-Darling Basin. A Report to the Australian Government from the CSIRO Murray-Darling Basin Sustainable Yields Project. Canberra, Australia: CSIRO. Retrieved 17 January 2014, from http://www.csiro.au/resources/WaterAvailabilityIn-Murray-DarlingBasinMDBSY

CSIRO (Commonwealth Scientific and Industrial Research Organisation). (2009). Water in Northern Australia. Summary of Reports to the Australian Government for the CSIRO Northern Australian Sustainable Yields Project Retrieved 23 December 2010, from http://www.csiro.au/files/files/ps7l.pdf

CSIRO (Commonwealth Scientific and Industrial Research Organisation). (2010). Climate Variability and Change in South-Eastern Australia. A Synthesis of Findings from Phase 1 of the South Eastern Australian Climate Initiative (SEACI) (pp. 36). Canberra, Australia: CSIRO Water for a Healthy Country Flagship. Retrieved 17 January 2014, from http://www.nccarf.edu.au/localgov/resources/climate-variability-and-change-south-eastern-australia-synthesis-findings-phase-1-south

Cullen, P. (2011). Adapting to water scarcity: a global challenge for the 21st century. In V. Cullen (Ed.), *This Land Our Water* (199-210). Hindmarsh, SA, Australia: ATF press.

Davies, P. E., Harris, J. H., Hillman, T. J. & Walker, K. F. (2010). The Sustainable Rivers Audit: assessing river ecosystem health in the Murray-Darling Basin, Australia. *Marine and Freshwater Research*, *61*, 764-777.

Denholm, M. (2010). Apple Isle's food bowl future, *The Australian 12 November 2010.* Retrieved 20 January 2014, from http://www.theaustralian.com.au/news/features/apple-isles-food-bowl-future/story-e6frg6z6-1225952283412

Dry, P. R., Loveys, B. R., Botting, D. & During, H.(1996). Effects of partial root-zone drying on grapevine vigour, yield, composition of fruit and use of water. *Proceedings of the 9th Australian Wine Industry Technical Conference*, 126 – 131.

Fairweather, H., Austin, N. & Hope, M. (Undated). Water Use Efficiency. An Information Package. Irrigation Insights No 5. Canberra, Australia: Land & Water Australia.

Finlayson, B., Peel, M. & McMahon T. A. (2011). Understanding global hydrology. In R. Quentin Grafton & K. Hussey (Eds.), *Water Resources Planning and Management*, Cambridge, UK: Cambridge University Press.

GHD. (2007). Using Recycled Water for Drinking, Waterlines Report Occasional Paper No 2. Canberra, Australia: National Water Commission.

Grabham, M., Hornbuckle, J., Raine, S. & Smith, R. (Undated). Observations from Initial Evaluation Trials of Bankless Channel Irrigation Systems: CRC for Irrigation Futures. Retrieved 17 January 2014, from http://irrigation.org.au/documents/publications-resources/conference_papers_2008/210508_Stream3_MGrabham1.pdf

Gregory, A. & Hall, M. (2011). Urban water sustainability. In I. Prosser (Ed.), *Water-Science and Solutions for Australia* (75-88). Canberra, Australia: CSIRO Water for a Healthy Country Flagship.

Hajkowitz, S. (2007). A comparison of multiple criteria analysis and unaided approaches to environmental decision making. *Environmental Science. and Policy*, *10*, 177-184.

Hall, A., Lamb, D. W., Holzapfel, B. & Louis, J. (2002). Optical remote sensing applications in viticulture - A review. *Australian Journal of Grape and Wine Research, 8*(1), 36-47.

Hamstead, M. (2007). *What is Environmental Water?* Paper presented at the 5th Australian Stream Management Conference. Australian Rivers: Making a Difference Thurgoona NSW, Australia.

Hatton MacDonald, D., Bark, R., MacRae, A., Kalivas, T., Grandgirard, A. & Strathearn, S. (2013). An interview methodology for exploring the values that community leaders assign to multiple-use landscapes. *Ecology and Society*, *18*(1), 29. http://dx.doi.org/ 10.5751/ES-05191-180129

Heaney, A., Beare, S. & Hafi, A. (2004). Trading with the evironment. *Australian Commodities*, *11*(4), 606-610.

Humphreys, E., Lewin, L. G., Khan, S., Beecher, H. G., Lacy, J. M., Thompson, J. A., et al. (2006). Integration of approaches to increasing water use efficiency in rice-based systems in southeast Australia. *Field Crops Research*, *97 (Special Issue)*, 19-33.

Humphreys, E., Meisner, C., Gupta, R., Timsina, J., Beecher, H. G., Tang, Y. L., et al. (2005). Water saving in rice-wheat systems. *Plant Production Science*, *8*(3), 242-258.

Humphreys, L., Muirhead, W., Lewin, L., Thompson, J., Stapper, M., O'Keefe, K., et al. (2003). Increasing water use efficiency. *Natural Resource Management*, June 2003 Special Issue, 28-35.

IACSEA (Independent Advisory Committee on Social and Economic Analysis). (1998). Socio-economic Assessment Guidelines for River, Groundwater and Water Management Committees. Sydney, Australia: NSW Department of Land and Water Conservation

IPCC (Intergovernmental Panel on Climate Change). (2013). *Summary for Policymakers. In: Climate Change 2013: The Physical Science Basis. Contribution of Working Group I to the Fifth Assessment Report of the Intergovernmental Panel on Climate Change.* Cambridge University Press, Cambridge, United Kingdom and New York, USA: IPCC

Jabro, J. D., Leib, B. G. & Jabro, A. D. (2005). Estimating soil water content using site-specific calibration of capacitance measurements from Sentek EnviroSCAN systems. *Applied Engineering in Agriculture, 21*(3), 393-399.

Jayawardane, N. S., Biswas, T. K., Blackwell, J. & Cook, F. J. (2001). Management of salinity and sodicity in a land FILTER system, for treating saline wastewater on a saline-sodic soil. *Australian Journal of Soil Research, 39*(6), 1247-1258.

Khan, S., Khan, M. A., Hanjra, M. A. & Mu, J. (2009). Pathways to reduce the environmental footprints of water and energy inputs in food production. *Food Policy, 34*(2), 141-149. doi: http://dx.doi.org/10.1016/j.foodpol.2008.11.002

Khan, S., Tariq, R., Yuanlai, C. & Blackwell, J. (2006). Can irrigation be sustainable? *Agricultural Water Management, 80*(1-3 (Special Issue)), 87-99.

Kirby, M. (2011). Irrigation. In I. Prosser (Ed.), *Water Science and Solutions for Australia* (105-118). Collingwood Victoria Australia.

Kirby, M. A. C., Qureshi, M. E., Mainuddin, M. & Dyack, B. (2006). Catchment behaviour and counter-cyclical water trade : an integrated model. *Natural Resource Modeling, 19*(4), 483-510. doi: 10.1111/j.1939-7445.2006.tb00191.x

Lamb, D. W. (2000). The use of qualitative airborne multispectral imaging for managing agricultural crops - A case study in south-eastern Australia. *Australian Journal of Experimental Agriculture, 40*(5), 725-738.

Lamb, D. W., Bramley, R. G. V. & Hall, A. (2004) Precision viticulture -an Australian perspective. *Acta Horticulturae, 640*, 15-25.

Lampard , J., Leusch, F. D. L., Roiko, A. & Chapman, H. F. (2010). Contaminants of concern in recycled water. *Water (Journal of the Australian Water Association), 37*(8), 54-59.

Lane-Miller, C. C., Wheeler, S., Bjornlund, H. & Connor, J. (2013). Acquiring water for the environment: lessons from natural resources management. *Journal of Environmental Policy and Planning*, *15*(4), 513-532.

Lane, M. B. & McDonald, G. (2005). Community-based environmental planning: Operational dilemmas, planning principles and possible remedies. *Journal of Environmental Planning and Management*, *48*(5), 709-731.

Lawford, R. (2011). Climate change and the global water cycle. In R. Quentin Grafton & K. Hussey (Eds.), *Water Resources Planning and Management*, (3-22). Cambridge, UK: Cambridge University Press.

Lukasiewicz, A., Finlayson, C. & Pittock, J. (2013). Identifying Low-Risk Climate Change Adaptation in Catchment Management While Avoiding Unintended Consequences. Gold Coast, Australia: National Climate Change Adaptation Research Facility. Retrieved 17 January 2014, from http://www.nccarf.edu.au/localgov/resources/climate-variability-and-change-south-eastern-australia-synthesis-findings-phase-1-south

Marsden, J. & Pickering, P. (2006). Securing Australia's Urban Water Supplies. Opportunties and Impediments (pp. 58). Camberwell, Vic, Australia: Marsden Jacob Associates for the Department of Prime Minister and Cabinet. Retrieved 17 January 2014, from http://www.environment.gov.au/water/publications/urban/securing-urban-water.html .

Marsden Jacob Associates. (2008). A National Snapshot of of Current and Planned Water Recycling and Reuse Rates, Report prepared for the Department of Environment, Water, Heritage and the Arts, Canberra, Australia. Retrieved 17 January 2014, from http://www.environment.gov.au/system/files/resources/86c2f1f4-2bf2-4923-a50d-986ace98e390/files/national-recycling-snapshot.pdf

McCartney, M. (2009). Living with dams: Managing the environmental impacts. *Water Policy*, *11*, 121-139.

McKenzie, F. & Williams, J. (2013). Australian Agriculture: Re-Designing for Resilience, Wentworth Group, Sydney, Australia. Retrieved 17 January 2014, from. http://www.wentworthgroup.org/docs/Australian_Agriculture_Redesigning_for_Resilience.pdf

MDBA (Murray-Darlin Basin Authority). (2010a). The Living Murray Environmental Water Recovery Progress Report. Canberra, Australia: MDBA. Retrieved 17 January 2014, from http://www.mdba.gov.au/files/publications/Assessing-environmental-water-needs-of-the-Basin-April-2010.pdf

MDBA (Murray-Darling Basin Authority). (2010b). The Basin Plan. Knowledge and Information Directory (BP-KID)Canberra, Australia: MDBA. (Map Figure 1 from http://www.mdba.gov.au/sites/default/files/cartographicmapping/163_Australian_Irrigation_Areas.pdf)

MDBA (Murray-Darling Basin Authority). (2012). Basin Plan. Retrieved 17 January 2014, from. http://www.mdba.gov.au/what-we-do/basin-plan

MDBA (Murray-Darling Basin Authority). (2013). Murray-Darling Basin Authority Annual Report 2012-13. Retrieved 17 January 2014, from http://www.mdba.gov.au/about-mdba/corporate-documents/annual-reports/ar-2012-13

Meyer, Wayne. (2003). Lift production by doubling yield from half the area. *Farming Ahead*, No *133*, January 2003, 54-55.

Meyer, W., Marvanek, S., Bryan, B., Christen, E., Hornbuckle, J., Khan, S., et al. (2005). Irrigation in Perspective: Irrigation in the Murray and Murrumbidgee Basins Canberra: CRC for Irrigation Futures. Retrieved 17 January 2014, from http://www.csiro.au/files/ files/p3we.pdf.

Meyer, W. & Montagu, K. (2006). The irrigation industry in the Murray and Murrumbidgee Basins. *Water (Australian Water Association)*, *33* (5), 53-56.

Meyer, W. S. & Bowmer, K. H. (2004). Sustainable irrigation. In K. H. Bowmer (Ed.), *Water Innovation: A New Era for Australia*, (104-125). Sydney, Australia: CL Creations.

Meyer, W.S. & Taylor. A.S. (1998). Audit and Strategy on Irrigation Education and Skills Development in Australia. LWRRDC Occasional Paper Series No. 21/98. Land and Water Resources Research and Development. Canberra, Australia. 54pp. Retrieved 6 February 2014, from http://lwa.gov.au/files/products/national-program-irrigation-research-and-developme/pr980297/pr980297.pdf

Murray Dairy. (2012). Annual Report 2012. Retrieved 17 January 2014, from http://www. murraydairy.com.au

Narayan, K. A., Schleeberger, C. & Bristow, K. L. (2007). Modelling seawater intrusion in the Burdekin Delta Irrigation Area, North Queensland, Australia. *Agricultural Water Management*, *89*(3), 217-228.

Newell, B., Crumley, C. L., Hassan, N., Lambin, E. F., Pahl-Wostl, C., Underdal, A., et al. (2005). A conceptual template for integrative human-environment research. *Global Environmental Change Part A*, *15*(4), 299-307.

NLWRA (National Land and Water Resources Audit). (2002). Australia's Natural Resources: 1997-2002 and Beyond, Canberra, Australia: NLWRA on behalf of the Commonwealth of Australia. Retrieved 17 January 2014, from http://www.anra.gov.au/topics/ publications/final-report/index.html

Nordblom, T. L., Christy, B. P., Finlayson, J. D., Roberts, A. M. & Kelly, J. A. (2010). Least cost landuse changes for targeted catchment salt load and water yield impacts in South Eastern Australia. *Agricultural Water Management*, *97*, 811-823.

Nordblom, T. L., Reeson, A. F., Finlayson, J. D., Hume, I. H., Whitten, S. M. & Kelly, J. A. (2011). Price discovery and distribution of water rights linking upstream tree plantations to downstream water markets: Experimental results. *Water Policy*, *13*(6), 810-827.

NSW Department of Primary Industries, Office of Water, Water for Rivers and State Water. (Undated). Water balance and unaccounted difference in the Murrumbidgee Regulated River. Retrieved 17 January 2014, from. http://www.carmproject.com.au/water-balance-murrumbidgee.pdf

NSW Government. *Water Sharing Plan for the Murrumbidgee Regulated Water Source 2003 Order under the Water Manangment Act 2000.* Retrieved 17 January 2014, from http://www.water.nsw.gov.au/Water-management/Water-sharing-plans/Plans-commenced/Water-source/Murrumbidgee-Regulated-River/default.aspx

NSW MWWG (New south Wales Murray Wetlands Working Group). (Undated). Adaptive Environmental Water Allocation. Retrieved 17 January 2014, from http://www.mwwg. org.au/allocation.phpW

NWC (National Water Commission). (2007). Australian Water Resources 2005. A Baseline Assessment of Water Resources for the National Water Initiative, Canberra, Australia: NWC. Retrieved 17 January 2014, from. http://www.water.gov.au/RegionalWater ResourcesAssessments/index.aspx?Menu=Level1_7

NWC (National Water Commission). (2010a). The Coal Seam Gas and Water Challenge: the National Water Commission Position, Canberra, Australia: NWC. Retrieved 17 January 2014, fromhttp://www.nwc.gov.au/ resources/documents/Coal_Seam_Gas.pdf.

NWC (National Water Commission). (2010b). Surface and/or Groundwater Interception Activities: Initial Estimates, Waterlines Report no 30, Canberra, Australia: NWC. Retrieved 17 January 2014, from http://www.nwc.gov.au/www/html/2826-surface-andor-groundwater-interception-activities---no-30.asp?intSiteID=1

NWC (National Water Commission). (2010c). The Impacts of Water Trading in the Southern Murray-Darling Basin. An Economic, Social and Environemental Assessment. Retrieved 17 January 2014, from http://www.nwc.gov.au/resources/documents/Pages_from_681NWC_ImpactsofTrade_ExecSummary-and_snapshot.pdf

NWC (National Water Commission). (2011a). National Water Commission - 2011 Stakeholder forum communique, 17 March 2011. Canberra, Australia: NWC Retrieved 17 January 2014, from http://www.nwc.gov.au/www/html/3016-2011-stakeholder-communique.asp?intSiteID=1

NWC (National Water Commission). (2011b). Strengthening Australia's Water Markets, Canberra, Australia: NWC.Retrieved 17 January 2014, from http://nwc.gov.au/__data/assets/pdf_file/0008/11240/StrengtheningAustraliasWaterMarketsReport.pdf

NWC (National Water Commission). (2012). National Water Planning Report Card 2011. Canberra, Australia: NWC. Retrieved 17 January 2014, from http://www.nwc.gov. au/?a=19805

NWC (National Water Commission). (2013a). Current Issues Affecting Australian Water Markets. Canberra, Australia: National Water Commission. Retrieved 17 January 2014, from http://nwc.gov.au/__data/assets/pdf_file/0018/35532/Current-issues.pdf.

NWC (National Water Commission). (2013b). National Water Account 2012, Canberra, Australia: NWC. Retrieved 17 January 2014, from http://www.bom.gov.au/water/nwa/ 2012/

NWC (National Water Commission). (2014). Appendix 3: Water Management Mechanisms used in Australia: Water Management and Pathways to Sustainable Levels of Extraction: Issues Paper Canberra, Australia: NWC Canberra, Australia: NWC. Retrieved 17 January 2014, from http://nwc.gov.au/publications/topic/water-planning/sle/appendix-c

Oakes, T. (Undated) Total Channel Control® - 10 Years Old and Growing Stronger. Retrieved 17 January 2014, from http://irrigation.org.au/documents/publicationsresources/conference_papers_2012/Experience_in_modernising_water_supply_and_irrigation_canals.pdf

OECD (Organisation for Economic Co-operation and Development). (2008). *OECD Environmental Performance Reviews Australia*. SourceOECD

Peterson, D., Dwyer, G., Appels, D. & Fry, J. (2004). Modelling Water Trade in the Southern Murray-Darling Basin. East Melbourne, Australia: Productivity Commission. Retrieved 17 January 2014, from http://www.pc.gov.au/__data/assets/pdf_file/0003/60474/ watertrade.pdf

Possingham, J. V. (2002). The influence of controlled water inputs on grape quality in regions of Australia with hot Mediterranean climates. *Acta Horticulturae*, *582*, 101-107.

Prathapar, S. A. (2000). Water shortages in the 21st century. In H. Cadman (Ed.), *The Food and Environment Tightrope.* (125-133). Canberra , Australia: Australian Centre for International Agricultural Research,.

Prathapar, S. A., Meyer, W. S., Madden, J. C. & Alociljá, E. (1997). SWAGMAN options: A hierarchical multicriteria framework to identify profitable land uses that minimize water table rise and salinization. *Applied Mathematics and Computation*, *83*(2-3), 217-240.

Proust, K., Dovers, S., Foran, B., Newell, B., Steffan, W. & Troy, P. (2007). Climate, Energy and Water: Accounting for the Links. Canberra, Australia Land & Water Australia. Retrieved 17 January 2014, from http://lwa.gov.au/files/products/land-and-water-australia-corporate/er071256/er071256.pdf

Purcell, J. & Currey, A. (2003). Gaining Acceptance of Water Use Efficiency Framework,Terms and Definitions, Final Report. National Program for Sustainable Irrigation. Retrieved 17 January 2014, from http://www.irrigationefficiency.co.nz/ assets/Uploads/NPSI.Au-9.pdf

Quentin Grafton, R. & Hussey, K. (2011). *Water Resources Planning and Management*. Cambridge, UK: Cambridge University Press.

Quentin Grafton, R. , Pittock, J., Davis, R., Williams, J., Fu, G., Warburton, M., et al. (2013). Global insights into water resources, climate change and governance. *Nature Climate Change*, *3*(4), 315-321.

Quiggan, J. (2012). Why the Guide to the Proposed Basin Plan failed, and what can be done to fix it. In J. Quiggin, T. Mallawaarachchi & S. Chambers (Eds.), *Water Policy Reform Lessons in Sustainability from the Murray-Darling Basin*, (49-60). Cheltenham, UK: Edward Elgar.

Qureshi, M. E., Shi, T., Qureshi, S. E. & Proctor, W. (2009). Removing barriers to facilitate efficient water markets in the Murray-Darling Basin of Australia. *Agricultural Water Management*, *96*(11), 1641-1651.

Race, D., Curtis, A., Birckhead, J., Boyd, M., Cooke, P., McDonald, S., et al. (2007). Assessment of irrigator's responses to total channel control and channel automation in the Central Goulburn irrigation area, Victoria. Report No 32 Albury, NSW, Australia: Institute for Land, Water and Society. Retrieved 28 January 2014, from http://137.166.4.200/research/ilws/research/publications/docs/report32.pdf

Radcliffe, J. C. (2003). Water - Second, Third Time Round *ATSE 2003 Symposium. Water - the Australian Dilemma* (211-219). Melbourne.

Australia: Australian Academy of Technological Sciences and Engineering.

Raine, S., Purcell, J. & Schmidt, E. (2005). *Improving whole farm and infield irrigation efficiencies using Irrimate TM tools.* Paper presented at the Irrigation 2005 Restoring the Balance, National Conference Irrigation Association of Australia, Townsville, Australia. Retrieved 17 January 2014, from https://eprints.usq.edu.au/565/1/Raine_etal_IAA2005_ImprovingWholeFarmandInfieldIrrigationEfficiencies.pdf

RGA (Ricegrowers Association). (2014). Environmental Champions Program Retrieved 17 January 2014, from http://www.rga.org.au/ecp-about.aspx

Richards, Q. D., Bange, M. P. & Johnston, S. B. (2008). HydroLOGIC: An irrigation management system for Australian cotton. *Agricultural Systems, 98*(1), 40-49.

Ross, H., Driml, S. & Zarezadeh, Z. (2012). Water Allocation, Social Change and Resilience. In J. Quiggin, T. Mallawaarachchi & S. Chambers (Eds.), *Water Policy Reform Lessons in Sustainability form the Murray-Darling Basin*, (179-192). Cheltenham, UK: Edward Elgar.

Roth, G., Harris, G., Gillies, M., Montgomery, J. & Wigginton, D. (2013). Water-use efficiency and productivity trends in Australian irrigated cotton: A review. *Crop and Pasture Science*, *64*(11-12), 1033-1048.

Ryan, S., Broderick, K., Sneddon, Y. & Andrews, K. (2010). Australia's NRM Governance System: Foundations and Principles for Meeting Future Challenges (pp. 87). Canberra, Australia: Australian NRM Chairs Committee. Retrieved 17 January 2014, from http://www.actnrmcouncil.org.au/files/NRM%20Governance.pdf.

Schofield, N. (2010). Australian wide assessment of river health. *Water (Journal of the Australian Water Association)*, *37*(1), 153-157.

SKM. (2005). State Water Operating Licence - Water balance template (pp. 34): IPART and Sinclair Knight Merz.

SKM et al. (Sinclair Knight Merz CSIRO and the Bureau of Rural Sciences). (2010). Surface and/or Groundwater Interception Activities: Initial Estimates *Waterlines report.* Canberra, Australia: National Water Commission. Retrieved 17 January 2014, from http://www.nwc.gov.au/resources/documents/Surface_and_or_groundwater_interception_activities_23June.pdf

Spies, B. & Dandy, G. (2012). *Sustainable Water Management. Securing Australia's Water in a Green Economy.* Melbourne, Victoria, Australia: Australian Academy of Technological Sciences and Engineering.

State Water. (2013). Progress on Cold Water Pollution, Retrieved 17 January 2014, from http://www.statewater.com.au/About%20us/News%20and%20events/Media%20releases%202012/Progress%20on%20cold%20water%20pollution%20project

Stirzaker, R., Lefroy, T., Keating, B. & Williams, J. (2000). A Revolution in Land Use. Emerging Land Use Systems for Managing Dryland Salinity Dickson, ACT, Australia: CSIRO

Stirzaker, R. J., Stevens, J. B., Annandale, J. G. & Steyn, J. M. (2010). Stages in the adoption of a wetting front detector. *Irrigation and Drainage, 59*(4), 367-376.

Syme, G. J. (2013). Acceptable risk and social values: struggling with uncertainty in Australian water allocation. *Stochastic Environmental Research and Risk Assessment*, *28*, 113-121.

Syme, G. J. & Nancarrow, B. E. (2008). Justice and the Allocation of Benefits from Water. *Social Alternatives*, *27*(3), 21-25.

Tan, P.-L., Mooney, C., White, I., Hoverman, S., Mackenzie, J., Burry, K., et al. (2010). Tools for Water Planning: Lessons, Gaps and Adoption *Waterlines Report No 37 National Water Commission* (pp. 388). Canberra, Australia: National Water Commision. Retrieved 17 January 2014, from http://nwc.gov.au/__data/assets/pdf_file/0003/10947/Waterlines_no37_tools.pdf

Tee, M., Burrows, D., Boland, A.-M. & Putland, S. (2003). Best Irrigation Management Practices for Vineyards in the Murray-Darling Basin. Glen Osmond, SA, Australia: Cooperative Research Centre for Viticulture. Retrieved 17 January 2014, from http://www.crcv.com.au/resources/Irrigation%20and%20Water/Additional%20Resources/MDB%20Irrigation%20Booklet.pdf

Troy, P. (2011). The urban water challenge in Australian cities. In Quentin Grafton, R., & Hussey, K (Eds.) *Water Resources Planning and Management* (463-482). Cambridge, UK: Cambridge University Press.

van Kalken, T. (2013). Improving River Efficiency. Modelling the Murrumbidgee to Improve River Management. DHI Group. Retrieved 17 January 2014, from http://www.dhigroup. com/upload/publications/scribd/168763102-Improving-River-Efficiency-DHI-Case-Story-AU.pdf

Vertessy, R. A., Zhang, L. & Dawes, W. R. (2003). Plantations, river flows and river salinity. *Australian Forestry*, *66*(1), 55-61.

Walker, B. & Salt, D. (2012). *Resilience Practice: Building Capacity to Absorb Disturbance and Maintain Function*. Washington DC , USA: Island Press

Watts, R. J., Allan, C., Bowmer, K. H., Page, K. J., Ryder, D. S. & Wilson, A. L. (2009). Pulsed Flows: A Review of Environmental Costs and Benefits and Best Practice. Canberra, Australia: *Waterlines*, National Water Commission. Retrieved 17 January 2014, from http://www.nwc.gov.au/resources/documents/Waterlines_report_No_16__Pulsed_flows_full_version.pdf

Wells, S., & McLean, J. (2013). One Way Forward to Beat the Newtonian Habit with a Complexity Perspective on Organisational Change. *Systems*, *1*, 66-84.

Wheeler, S., Loch, A., Zuo, A. & Bjornlund, H. (2013). Reviewing the adoption and impact of water markets in the Murray-Darling Basin, Australia. *Journal of Hydrology*.

Williams, J., Milligan, A. & Stubbs, T. (2013). Coal Seam Gas Production: Challenges and Opportunities. Canberra, Australia: Bureau of Resources and Energy Economics. Retrieved 17 January 2014, from. http://wentworthgroup.org/wp-content/uploads/2013/ 12/BREE_Coal-seam-gas-production_WILLIAMS-etal-.pdf.

Wittwer, G. & Dixon, J. (2013). Effective use of public funding in the Murray-Darling Basin: a comparison of buybacks and infrastructure upgrades. *Australian Journal of Agricultural and Resource Economics, 57*(3), 399-421. doi: 10.1111/1467-8489.12001

Wong, T. & Brown, R. (2011). Water sensitive urban design. In R. Quentin Grafton & K. Hussey (Eds.), *Water Resources Planning and Management* (483-504). Cambridge, UK: Cambridge University Press.

Wood, M. L. & Finger, L. (2006). Influence of irrigation method on water use and production of perennial pastures in northern Victoria. *Australian Journal of Experimental Agriculture*, *46*(12), 1605-1614.

Xevi, E. & Khan, S. (2003). *A Multi-Objective Optimization Approach to Water Management.* Paper presented at MODSIM, Townsville, Australia.

Young, W. J. & Chiew, F. H. S. (2011). Climate change in the Murray-Darling Baisn: implications for water use and environmental consequences. In Quentin Grafton R & K. Hussey (Eds.), *Water Resources Planning and Management*, (439-460). Cambridge, UK: Cambridge University press.

In: Drinking Water and Water Management
Editor: Mason Giannino
ISBN: 978-1-63117-523-7

Chapter 8

RESILIENT IRRIGATION THROUGH WATER TRADE: ISSUES, INVESTIGATION AND INSIGHTS

Adam Loch[*]
Centre for Regulation and Market Analysis, Business School, University of South Australia, Adelaid, South Australia

ABSTRACT

The idea of resilience captures individual, sectoral or community capacity to tackle disruptive events that change their status quo, and to successfully transform themselves into new states of existence. During the Millennium drought in Australia, many irrigators faced severe water shortages and unusual government market activity which constituted significant disruptive events. Coincidentally, water reform prior to these events provided scope for water trading that, together with other adjustment packages and support, arguably offered a major method for achieving resilience among affected irrigator groups. This chapter reviews those events and the resilience outcomes from water trade activity in Australia's Murray-Darling Basin during the Millennium drought, with an objective of attempting to tangibly link irrigator resilience to water trade availability in the Basin. A methodological approach for testing irrigator resilience outcomes is proposed and specified for subsequent modelling using irrigator water trade data

[*] Phone: +61 8 8302 7296. Email: adam.loch@unisa.edu.au

collected during the Millennium drought. Knowledge of the properties that confer resilience in irrigation systems may enable other resource managers, users and communities to design and implement appropriate polices that minimize social impacts while maximizing sustainable water management and ecosystem goods and services outcomes.

Keywords: resilience, water markets, environmental water, Murray-Darling Basin

INTRODUCTION

Most irrigation districts were established globally in a period when we had far less appreciation of water resource limits and environmental requirements. These irrigation districts consequently formed a basis for rural communities to expand and prosper, as well as providing broader perceived economic benefits at state and national levels. Such global perceptions about the benefits of irrigation mean that it now accounts for 80% of fresh water use in the US (Wiebe and Gollehon, 2006), while the World Bank (1992) estimates that at least 70% of fresh water is used globally for irrigated agriculture. To support access to, and reliability in the supply of, fresh water resources for irrigation many countries have invested heavily in fixed infrastructure to store, deliver and reuse water resources. While the economics of such irrigation investment have been questioned by some (e.g., Davidson, 1969), generally such concerns are dismissed by planners and policy-makers eager to tame river systems for social and economic security (Griffin, 2006). However, over time the environmental impacts of such high irrigation water use and supporting infrastructure investment have become apparent in the form of natural flow reductions, land and water salinity increases, and algal bloom outbreaks (Leblanc et al., 2012, Blackmore, 2001).

As recognition of environmental water requirements increased and the notion of sustainable resource use took root, a shift in public thinking also emerged. At governance levels policy-makers were faced with significant trade-offs to implement and manage, while irrigators and rural communities faced significant change, challenges and uncertainty. Initial arrangements to address sustainable water requirements focused on protection and control policies—as well as regulatory attempts to prioritize water for environmental use where no surplus stocks existed (Loch et al., 2011). Such polices can, however, inadvertently compromise the ability of resource users and communities to adapt and be resilient (Marshall et al., 2007). Therefore, many countries adopted market-based instruments as a basis for reallocating water between different uses (i.e., economic, social and environmental water requirements), particularly tradable permit systems (Jordan et al., 2005) such as property right reforms and water markets. Despite this clearly perceived link between water trade and increased social resilience outcomes, a structured assessment of how the implementation of water market has specifically impacted on irrigator and social resilience has not been undertaken.

But resilience is relevant to more than the social context. For researchers, a priority is to understand the social *and* ecological consequences of market-based water reforms (Connor et al., 2013). Such analysis can provide an understanding of whether these market-based reforms complement or clash with irrigation and rural community interests in sustainable water resource management—which poses a significant challenge for irrigation districts around the world. Since resilience is a concept that can be used to guide and support more effective socio-ecological management systems and institutions (Berkes and Folke, 1998), as well as guiding transformation processes required to promote sustainable and/or innovative rural outcomes (Ross et al., 2012), this chapter will examine issues associated with resilience and water markets to propose a means of potentially assessing their usefulness in helping irrigators address such challenges.

For context, the Murray-Darling Basin (MDB) in Australia's south-east will be used (Figure 1), where irrigators and other water users recently experienced a prolonged period of extreme drought and water scarcity. This drought required water users to adapt and, in some instances, transform their operations dramatically in order to cope. Much of this adaptation was conducted through water-markets. As such, the MDB provides a useful basis for framing and examining resilience outcomes under water trade arrangements. Knowledge of the properties that confer resilience in irrigation systems may enable other resource managers, users and communities to design and implement appropriate polices that minimize social impacts while maximizing sustainable ecosystem goods and services (Adger, 2000, Crean, 1999). We begin with a consideration of the resilience concept in relation to socio-ecological water management.

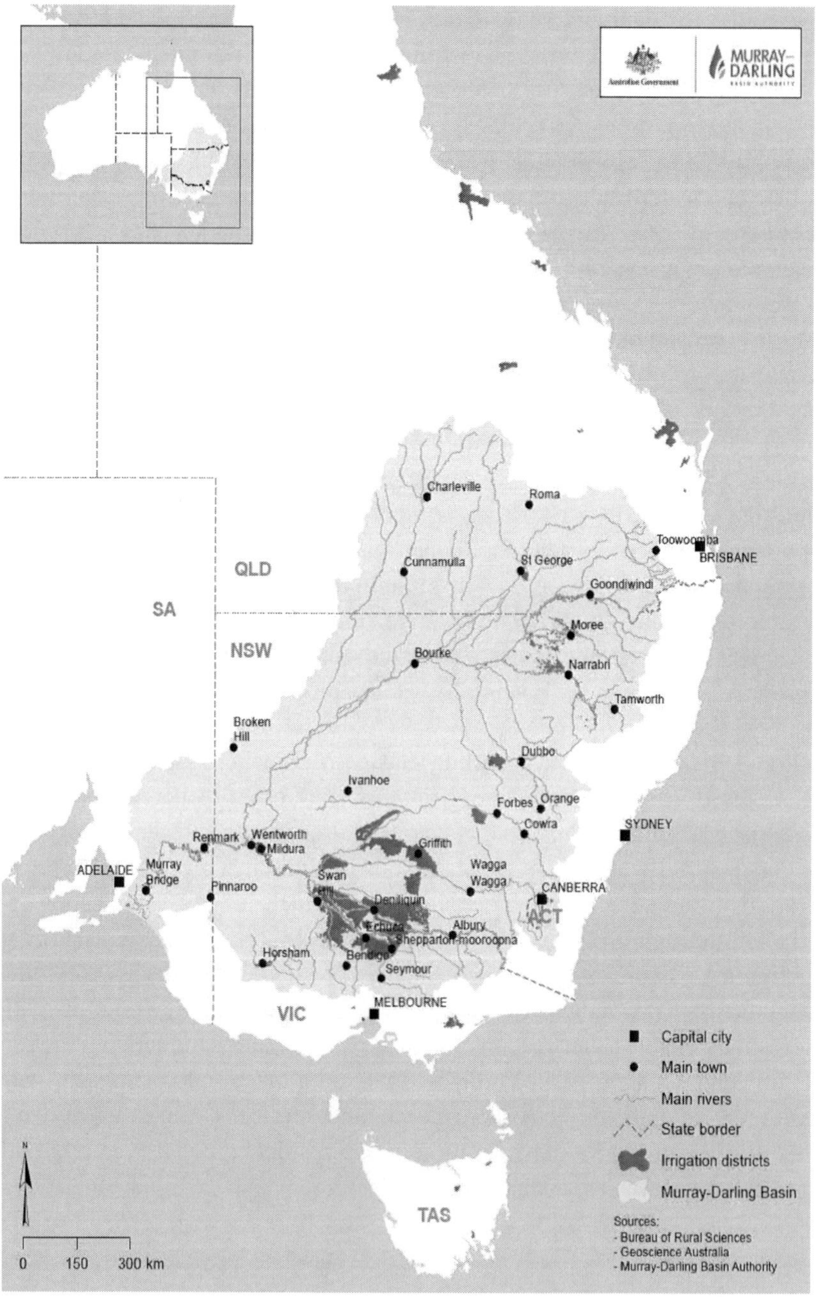

Figure 1. The Murray-Darling Basin (MDBA, 2012).

Resilience in Water Management

Resilience is broadly defined as a system or groups' capacity to absorb shocks and maintain its normal functioning or identity. However, this definition can be mainly prescribed toward social resilience which is the ability of communities to absorb external changes and stresses while maintaining the sustainability of their livelihoods (Adger et al., 2002). Social resilience may also concern a community's capacity for renewal, reorganization and transformation in the face of stressors, rather than maintaining a status quo (Gunderson and Holling, 2002), chiefly through innovation and the development of strategies essential for sustainability (Folke, 2006). This expansion of the resilience definition provides scope for different approaches to identifying resilience arrangements in social settings, which can be considered more reflective of tangible social activity. By way of example, Australian irrigated agriculture has transformed itself repeatedly over its history within a series of exploratory, expansion, maturity and increasing scarcity, and finally contraction phases in response to an emphasis on sustainable water resource management and use (Figure 2). The most recent contraction phase of transformation has required Australia's irrigated

Market characteristic	Exploration	Expansion	Maturity	Contraction	
Long run supply	Low, but growing	Low, rapidly growing	High and growing	High, but stable	
Water demand	Elastic, often low or zero prices	Elastic low prices	Elastic at low prices; inelastic at high prices	Elastic at low prices; no longer perfectly inelastic at high prices	
Water competition	Nil	Minimal	Intense	Easing somewhat	
Externalities	Nil	Minimal	Pressing	Reducing	
Water market development	Origins of water markets		Emergence of markets	Development of markets	Sustainable water trade
Timeline	European settlement	1900-1980	1980-1994	1994-2004	2004-present

Figure 2. Phases of Australian irrigation transition (Cummins and Watson, 2012, NWC, 2011).

agriculture sector to adapt to (among other issues) water scarcity pressure via the institution of competitive markets for water resource reallocation. In turn, the sector has been forced to grasp and respond to the opportunity costs of water resource use through forms of progressive innovation and technological adoption (Mallawaarachchi and Harris, in press).

However, while societies may exhibit significant capacity to adapt and transform to disturbances when analyzed through a social lens, such adaptation may be at the expense of changes in the capacity of ecological systems to absorb change pressures (Smit and Wandel, 2006). These system disparities may generate breakpoints in the total social-ecological resilience model (Gunderson and Holling, 2002), suggesting social resilience may not always be an optimal goal (Folke, 2006), and actually result in vulnerable systems overall. Vulnerable social-ecological systems imply a loss of adaptability, which is the capacity of agents in a system to build resilience through collective action. Adaptability is thus arguably different from transformation when viewed as an agents' capacity to create fundamentally new systems where ecological, political, social or economic conditions make the status quo untenable (Walker et al., 2004). A research challenge is thus how to design (identify) policies that stimulate: i) adaptive governance for resilience toward short-term conditions; and ii) transformations that can cope with sustainable outcome requirements in the distant future (Folke, 2006). An examination of the Australian water market policy context, and how irrigators used such institutions to adapt during the Millennium drought (1998 to 2010), may be helpful in determining answers to this research challenge.

CHALLENGES IN SOCIO-ECOLOGICAL RESILIENCE POLICY SETTING

Policies at the national, state and local governance levels all impact upon systems and their capacity to be resilient (Ross et al., 2012). In Australia a cross-governance recognition of the need to direct social, economic and environmental water requirements onto a sustainable trajectory was formally reached in the 1990s (ARMCANZ, 1996). However, the paradox between water resource protection and control (externality management) requirements and socio-economic adaptation (freedom in social, economic and environmental choice) requirements often resulted in difficult policy trade-offs at national, state and local governance levels. The period that followed (until

2004) exemplified a range of issues which threatened to derail the reform process and result in poor water management and socio-ecological resilience outcomes. These issues included:

- Political tensions arising from state ownership and control of water resources requiring transfer payment withholding strategies by the national government to motivate prioritisation of sustainable water management objectives;
- Differences in state legislation outcomes, water management planning approaches, data collection, modelling analysis approaches, terminology, and property right specification arrangements—as well as some catchment boundaries overlapping different jurisdictional contexts—which created barriers to efficient and effective water management consistent with national goals(Loch et al., accepted);
- Irrigator uncertainty and concerns about asset values and future investment as water and land property rights were unbundled (separated) in readiness for trade, and water entitlement right shares were altered toward volumetric shares of a consumptive resource specified in (emerging over this period) catchment water management plans;
- Community concerns regarding possible large transfers of water away from rural districts as a consequence of water trade availability, which was perceived to result in decreased future employment opportunities, economic decline and general population emigration and drive social vulnerability(Loch et al., 2013, NWC, 2011); and
- Poor comprehension of sustainable water resource requirements at various scales (local, state and national), as well as significant debate about what constituted appropriate sustainable water management outcomes. This highlighted possible ecological vulnerability and the need for tangible environmental representation in the debate.

These disputes collectively stalled the MDB reform process toward sustainable water management over a period of 10 years. In that time, irrigators and irrigation communities suffered a high degree of future business and regulatory uncertainty, while in addition no significant volume of water was transferred toward meeting sustainable environmental use objectives. Even with increased access to water markets over this timeline, few irrigators used the market toward adaptive capacity or resilience ends. Further, with its national sustainable water reform agenda threatened, the federal government enacted a new policy approach in an attempt to bring traction back to the process.

A REVISED MDB POLICY APPROACH

In 2004 a *National Water Initiative* agreed to by all government parties was implemented in order to address the problems raised above. Under a further agreement specific to the MDB, $500 million was made available to recover 500 GL of water for the environment through: i) investments in irrigation infrastructure to deliver water savings that would be shared between users and the environment; and ii)water purchases through the now rapidly maturing market institutions (Adamson and Loch, in press).[1] The amount of money was later doubled to $1 billion, and a revised recovery target set for 2009 (Loch et al., 2013). The process began reasonably well, with some tangible recovery of environmental water being accomplished for the first time. However, by 2006/07 the cumulative diminished inflow effects of the Millennium drought on MDB water storages reduced opening irrigation water allocation announcements to unprecedented low levels(Wheeler et al., 2012). This prompted the Prime Minister to act unilaterally in 2006 and announce a new *National Plan for Water Security.* This policy program (with a predominant focus on MDB objectives and sites) allocated a further $10 billion toward the implementation of infrastructure efficiency improvement and water market recovery initiatives to satisfy new planning objectives—this time aimed at recovering 1500 GL of water for MDB environmental use.

Although there was a change of government in 2007, the national water recovery plan remained in place and was further bolstered by an additional $2.9 billion in funding to achieve its aims. At a time of unprecedented low irrigation water allocations, significant reductions in commodity returns and agricultural balance of trade positions, increased farm operating costs and uncertainty about farm succession benefits or arrangements many irrigators elected to sell part or all of their water rights to the federal government (Wheeler and Cheesman, 2013). Those who elected to continue farming during this drought period experienced some of the highest seasonal allocation water prices ever seen, intense competition for scarce water resources, significant movements of water from New South Wales (NSW) into Victoria (VIC) and South Australia (SA), and scarcity-induced unique market institutions; all of

[1] A gigalitre (GL) is roughly equivalent to 811 acre feet of water.

which had significant impacts on the volume of water trades and irrigator trade behaviour (Loch et al., 2012).Consequently, as at 30 September 2013 the water purchasing component of the national plan had recovered 1337 GL of water for the environment (DSEWPC, 2013).

It is generally agreed that the presence of water markets in the MDB during the Millennium drought period resulted in lower regional job losses (Wittwer and Griffith, 2011) and higher farm production outputs and profits (Dixon et al., 2011, DAFF, 2011, NWC, 2012) than would have been experienced otherwise—and obviously a significant volume of water has been recovered for sustainable use. However, how do we evaluate the success of water markets with regard to socio-ecological resilience outcomes, as defined at the beginning of this section? Can we conclude that water markets provide an appropriate policy instrument in support of irrigator resilience, at least in the context of the MDB? We consider these questions in the next section.

Water Markets and Resilience

To begin, it is important to note that a focus on water market drivers of socio-ecological resilience in this chapter, rather than a consideration of both market and infrastructure investment policies for sustainable resource reform, is based on an acknowledgement that the MDB infrastructure policy deliverables are long-run in nature and highly uncertain at this point in time (Loch et al., 2014).

Conversely, data on MDB irrigator decision-making and (partial) resilience outcomes is short-run and relatively more readily available through a unique dataset held at the University of South Australia (see Wheeler et al., 2012 for details). This data was compiled through a survey of 946 MDB irrigators about their past water trade and farm management decisions toward the end of the Millennium drought.

Water Trade Activity and Drivers

There are many examples of irrigator adaptation and transformation through access to, and use of, water markets during the Millennium drought period that can easily be considered a significant disturbance to the status quo.

The first involves a broad examination of water market uptake both prior to and following the drought disturbance (Figure 3).

As shown, seasonal allocation water trade is largely non-existent prior to the mid-1990s when broad agreement about the need for sustainable water reallocation is agreed to and a water resource use cap is implemented in the MDB, limiting further water extractions to satisfy new water right demands by consumptive users (MDBC, 2000). The cap stimulates a growth in trade activity, as demand must now be satisfied by transfers between users. In support of this, institutional measures to separate property rights and facilitate water trading had developed in the interim.

By 2000/01 continuing dry conditions drive further growth in allocation trade activity which increases exponentially after the 2006/07 allocation shocks (Figure 4) and a growing realization that storage levels are unlikely to be replenished by rain in the short-term.

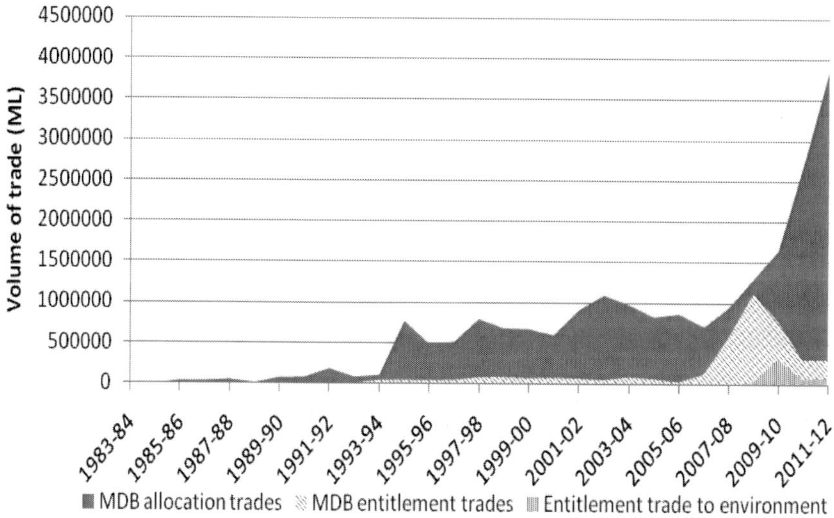

Figure 3. MDB allocation and entitlement water trade volume 1983-2010 (NWC, 2013).

This realization also appears to drive significant trade in water entitlement rights during the same period, with returns to relatively low levels of transfer at the conclusion of the drought disturbance (2009/10). The chart also records total environmental transfers of entitlement purchases through the market until a 2011 political shift forced a rethink on the use of water markets as a major policy program to recover sustainable water reserves (Australian Parliament, 2011). Despite this change in policy priority, water markets provided the means to quickly (relative to progress over the previous twelve-years) and effectively transfer significant volumes of water from consumptive (irrigation) to environmental ownership, thereby bolstering ecological resources and the probability of future sustainable resilience. Although water markets are still being employed to recover water, the current policy emphasis is now on infrastructure efficiency savings with arguably significantly higher cost requirements (Loch et al., 2014).

Such water trade behaviour by irrigators is also indicative of social resilience efforts through their desire to adapt to disturbance, exploring new ways to achieve adaptation requirements, and attaining a new equilibrium (resilience) in response to a prolonged period of variable water supply and disturbance. It is worth noting that individual irrigation farms might also have achieved resilience through the adoption of new technologies (practices) to provide flexibility in managing variable water supply conditions (Mallawaarachchi et al., 2010).[2] However, while an adoption of on-farm efficiency practice changes featured during this drought period (Loch et al., 2012), in the main irrigators used water trade activity to achieve partial farm transformation to new states of existence via reduction of existing water assets through sale to the government. Alternative forms of farm adaptation may involve changes to the enterprise mix, operational structure, farm location or business status—including bankruptcy or exit—which also featured in irrigator decision-making at this time (Loch et al., 2012).

Such extreme forms of farm transformation occurred during the drought where irrigators elected to exit completely from the agricultural sector and transform totally to a new state of existence elsewhere.

Those irrigators who did not sell their entitlement rights to the government and continued farming during the drought typically used seasonal allocation water trade as a reactive risk mitigation strategy in order to achieve production, income and farm viability resilience in response to high volatility (Mallawaarachchi and Harris, in press). Novel institutional water trade

[2] Adamson and Loch (in press) discuss the perverse outcomes from following such strategies at broader levels, with model estimates showing significant reductions in water availability to the environment over long-run timeframes as well as increased capital investment vulnerability for irrigators.

	High reliability entitlements						Lower reliability entitlements		
Year	Vic Goulburn	Vic Murray	NSW Murray	NSW Murrumbidgee	SA Murray	Vic Goulburn (low)	Vic Murray (low)	NSW Murray (general)	NSW Murrumbidgee (general)
1998-99	100%	100%	100%	100%	100%	0%	100%	93%	85%
1999-00	100%	100%	100%	100%	100%	0%	90%	35%	78%
2000-01	100%	100%	100%	100%	100%	0%	100%	95%	90%
2001-02	100%	100%	100%	100%	100%	0%	100%	105%	72%
2002-03	57%	100%	100%	100%	100%	0%	29%	10%	38%
2003-04	100%	100%	100%	95%	95%	0%	0%	55%	41%
2004-05	100%	100%	97%	95%	95%	0%	0%	49%	40%
2005-06	100%	100%	97%	95%	100%	0%	0%	63%	54%
2006-07	29%	95%	69%	90%	60%	0%	0%	0%	10%
2007-08	57%	43%	50%	90%	32%	0%	0%	0%	13%
2008-09	33%	35%	95%	95%	18%	0%	0%	9%	21%
2009-10	71%	100%	97%	95%	62%	0%	0%	27%	27%
2009-10	71%	100%	97%	95%	62%	0%	0%	27%	27%
2010-11	100%	100%	100%	100%	100%	0%	0%	100%	100%
LTAAY*	95%	95%	95%	95%	90%	35%	24%	81%	64%

Figure 4. MDB opening allocation levels 19998 to 2010 (NWC, 2013).

* Long-term average annual yield (LTAAY).

arrangements (e.g., limits to out of district transfers combined with ballot-draws to determine inclusion if the limits were exceeded) were also trialed in various irrigation districts to moderate trade behaviour and to provide proactive options for irrigators to use in adaptation to high supply variability expectations (Loch et al., 2012). For many irrigators this period was characterized by steep adaptation learning-curves in response to exceptionally high allocation water prices (e.g., up to AU$1100/ML, where normal allocation water prices may be around AU$50/ML), and difficult trade-offs related to the retirement of under-performing cropping areas which in many instances were pulled out and destroyed. Thus, allocation variability was a major driver of irrigator adaptation and transformation behaviour in the MDB through water markets (Figure 5). In general, it appeared that irrigators were willing to bear around 40-50% reductions in allocation volumes before substantial adaptation was instigated through the water markets in 2007/08.

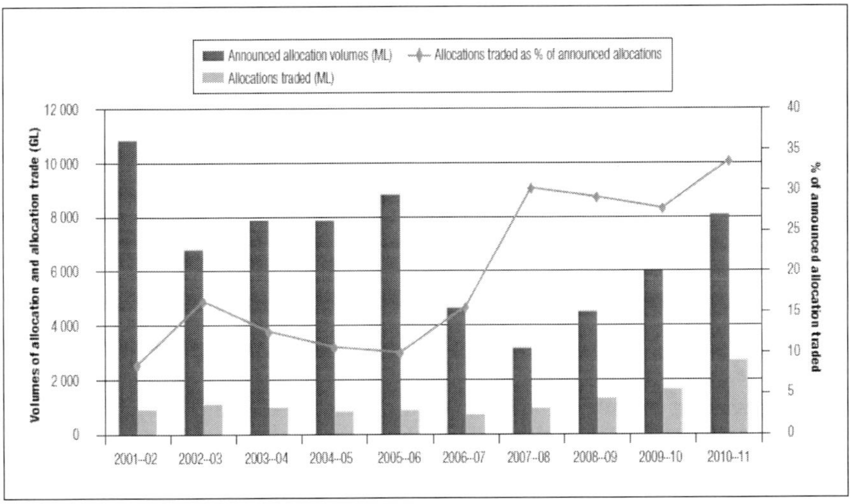

Figure 5. MDB allocation water trade volume 2001-2011 (NWC, 2013).

Socio-Ecological Insights from Water Market Assessments

How then might the question raised above—that is, the need to identify and/or design policies aimed at delivering short- and long-term adaptive and sustainable outcomes toward resilient socio-ecological systems—be enhanced by a more thorough investigation of water market institutions? It is clear from the previous discussion that water markets in the MDB have delivered significant sustainable water resource transfers, which will be used in future to derive ecological refugia, site-connectivity, flushing functions and water flow patterns to mimic wetting/drying periods across MDB environmental systems (Connor et al., 2013, Overton et al., 2009). Further, water markets have allowed irrigators to adapt to this sustainability requirement—even during a period of extreme disturbance in water supply conditions and future uncertainty. Studies into the potential impacts of government water trade to bolster environmental water applications in the MDB have also generally concluded that ecological benefits could be significantly enhanced without disadvantaging irrigators or rural communities (Kirby et al., 2006, Tisdell, 2010, Wheeler et al., 2013, Connor et al., 2013). As such, water markets appear to offer a useful approach to socio-ecological adaptation and resilience, which has not been investigated previously.

Prior studies into resource user resilience and adaptation tend to focus on the social aspects in isolation in order to examine policy lessons and outcomes. For example, Marshall et al. (2007) use resource dependency to measure and explore resource users' capacity to adapt and be resilient to changes in their situation (in this case commercial fishers in North Queensland). The concept of resource dependency is defined as a unique relationship between resource users and a resource (e.g., irrigation communities and water) such that changes in the nature of this relationship may affect the capacity of social and ecological systems to be resilient (Adger, 2000, Krannich and Zollinger, 1997, Bailey and Pomeroy, 1996). Water trade activity during the drought suggests varying levels of resource dependency by irrigators toward secure and reliable access to water resources, particularly during periods of scarcity. As such, resource dependency approaches could provide a useful basis for examining water market effects on adaptive capacity and resilience.

In addition, enhanced knowledge about resilience properties may assist us to minimize social impacts and maximize environmental sustainability outcomes from policy choices (Adger, 2000). Such an objective might be achieved by tracing the effects of institutional water trade approaches on irrigators' reactions to these new opportunities or constraints (Marsden Jacob Associates et al., 2010). Marshall et al. (2007) support this view where they argue that resilience can be influenced by a variety of factors, and that an understanding of resource dependency helps us assess why some individuals are more resilient than others. Therefore, identifying what influences adaptation and resilience through water markets may provide a means of analyzing the usefulness of such policy approaches in the MDB and other similar jurisdictions.

A Possible Methodology Framework

A suggested methodology for examining the relationship between water market policy and social resilience outcomes for irrigators is therefore proposed below.

Resource Dependency Measure and Variables

An initial issue is how to conceptualize and measure resource dependency in the MDB irrigation and water market setting. To quantify and examine the relationship between resource dependency and social resilience, Marshall et al. (2007) employed principle component analysis techniques using ten possible social, economic and environmental influences. These influences include: occupational attachment; attachment to place; employability; family attitudes to change; business size; business approach; financial situation; level of specialization; time spent harvesting; and interest in and knowledge of the environment. Although they don't specify a direct measure for adaptation or resource dependency, Mallawaarachchi and Harris (in press) suggest similar variables are related to irrigator resilience and flexibility from an analysis of MDB farming performance ABARES data over several years.[1] In particular, variables such as enterprise mix, annual versus perennial cropping choices, diversification, off-farm income, debt levels, farm planning and water use, application rates and trade at the individual farm level were determined by the authors to be important factors influencing irrigator sensitivity to drought shocks, as well as their adaptive capacity. Usefully, the University of South Australia dataset provides a range of variables (Table 1) that could be employed to construct a similar measure of resource dependency for MDB irrigators, and to identify important social resilience factors.

[1] Australian Bureau of Agricultural and Resource Economics and Science (ABARES).

Water Market Role in Adaptation

Linking possible influential variables to resource dependency among MDB irrigators may provide important insights into the role that water trade plays in adaptive capacity and resilience outcomes at both social and ecological levels. Interestingly, the findings of Marshall et al. (2007) provide early support for the importance of water market institutions in providing adaptive capacity. In that study, commercial fishers were found to be more vulnerable to proposed or possible institutional change due to their lack of skills, attitudes and opportunities to successfully navigate through a disturbance period. The commercial fishers were constrained in their capacity to experiment with new ideas, and less likely to take risks. As such, they were less likely to transform outside the industry (transformative adaptation), or remain viable within the industry (incremental adaptation).

Table 1. Component to variable matches for proposed methodology

Commercial fishing	Irrigation and water trade
Occupational attachment	Profit or farming attitudes
Attachment to place	Lifestyle attitudes, membership
Employability	Age, education, off-farm employment
Family attitude to change	Succession plan, family attitudes
Business size	Farm size, irrigated hectares
Business approach	Whole of farm planning, land/water sales
Financial situation	Debt levels, income
Level of specialisation	Efficiency investment, farm diversity
Time spent harvesting	Years spent farming
Interest in and knowledge of the environment	Environment attitudes, group membership

As outlined above, these same constraints do not appear to have manifest among MDB irrigators during the drought period due to broad access to water markets, within which irrigators could experiment using different adaptation strategies.

Consequently, MDB irrigators adopted a variety of water trade behaviour in response to the drought disturbance, reducing their vulnerability. Specifically, irrigators engaging only in seasonal allocation trade adopted an incremental adaptation strategy, possibly due to lower resource dependency.

On the other hand, irrigators electing to sell part or all of their entitlement rights to the government because of perceived higher resource dependency adopted a transformative adaptation strategy to cope with the disturbance stressors. Finally, irrigators engaging in no trade (which constituted a significant proportion of the dataset) could be considered to have no resource dependency (vulnerability) or requirement to adapt through water markets, therefore relying on other strategies to achieve or maintain resilience. Thus the freedom by MDB irrigators to engage (or not) with water markets provided an individualistic approach to risk management rather than imposing a 'one size fits all' policy option. Such flexibility matched well with the degree of heterogeneity typically found amongst resource dependent groups—the adaptive management of which typically requires simple, flexible and assisted programs to enable and facilitate changes in resource use (Marshall et al., 2007).

The task beyond this chapter is to adapt the approach of Marshall et al. in order to identify important influences of water trade across both seasonal allocation and entitlement transfer activity, as measures of social and ecological system adaptation, which might be used by policy-makers to identify or (re)design their approaches to combined system resilience objectives.

Usefulness of the Proposed Approach

For irrigators in other contexts facing similar sustainable water management pressures determining influences on irrigator/water market interaction may help policy-makers to better plan institutional designs, and assist them to prepare incentives for implementation during (likely) subsequent disturbance events. Such program planning may provide both risk management arrangements for users, as well as positive outcomes for sustainable water resource use through opportunistic arrangements. It may also be interesting to see whether the resilience component issues identified by Marshall et al. (2007) for vulnerable commercial fishers relate to resource dependency issues among MDB irrigators, or if different drivers are responsible for resilience outcomes in that context. Where similar components exist, adaptation and resilience objectives may arguably be met regardless of sectoral resource dependency (vulnerability).

In addition, for the Australian context additional light may be shed on the drivers of water scarcity adaptation that could be useful for future climate change management and policy improvement at the margin. As outlined above, the ability of water markets to allow irrigator experimentation with different strategic approaches during the drought was likely a critical adaptive capacity function. As such, continual assessment and novelty in water markets may be an issue requiring careful attention between disturbances such that all users, including the environment, will benefit from increased resilience across socio-ecological systems.

CONCLUSION

This chapter has illustrated how irrigators in Australia's MDB used water markets during a time of extreme drought disturbance to adapt to that change, manage risk and successfully address potential vulnerability. However, what lessons can be learned from this market-driven adaptation process to assist other resource users facing similar sustainability requirements and/or scarcity stresses? Further, what feedbacks for institutional reform at the margin can be applied from such insight to drive improvement in future public interventions to promote private and public water sustainability benefits?

We have seen that resilience among resource users is far more than simply coping with a disturbance event, and then returning to the original status quo position. It may involve potentially innovative adaptations that completely alter the original state of existence, allowing the systems to evolve, survive and continue in a sustainable new form. For socio-ecological resilience in particular, sustainable resource use issues may be complicated under combined resource-dependent systems with divergent interests, such as those found in the Murray-Darling Basin in Australia. This chapter proposes a link between water market institutions and resilience assessment approaches to assess how such economic concepts support socio-ecological outcomes, and fit with the notions of both static and transformative adaptation towards resilience. Further, the concept of resource dependency is identified as a possible measure for socio-ecological adaptive capacity, and proposed variables influencing that outcome are suggested for further testing.

Finally, a unique dataset of irrigator decision-making in the MDB collected during a significant drought disturbance event may provide a possible means to examining and answering questions related to the identification and (re)design of policy programs and initiatives to drive short- and long-term adaptive sustainable outcomes in both systems. If successful, the proposed approach may also strengthen previously unidentified links between water markets and socio-ecological resilience outcomes.

The next step will be to develop the proposed model specification and test the resource dependency metric further, including appropriate relationship properties (e.g., possible non-linear relationship properties between the variables). In addition, possible estimation techniques may have to include Heckman models where non-random sample selection techniques are used to derive observations. Non-random selection approaches result in subsequent biased model estimates, which Heckman estimation techniques can correct. However, the potential for policy insight from this approach is significant; especially among countries where water market institutions are still developing or where expected future disturbance effects on water resources are anticipated to result in large vulnerability consequences for resource users. As such, the proposed approach offers substantial benefits for irrigators and irrigation communities facing new or modified challenges from sustainable water management requirements in jurisdictions around the world.

REFERENCES

Adamson, D. & Loch, A. in press. Possible negative feedbacks from 'gold-plating' irrigation infrastructure. *Agricultural Water Management.*

Adger, W. N. 2000. Social and ecological resilience: Are they related? *Progress in Human Geography,* 24, 347-364.

Adger, W. N., Kelly, P. M., Winkels, A., Huy, L. Q. & Locke, C. 2002. Migration, remittances, livelihood trajectories, and social resilience. *AMBIO: A Journal of the Human Environment,* 31, 358-366.

Armcanz 1996. National principles for the provision of water for ecosystems. *Occasional paper SWR No. 3, July.* Canberra: Dept. of the Environment, Sport and Territories.

Australian Parliament 2011. Of drought and flooding rains: Inquiry into the impact of the Guide to the Murray-Darling Basin Plan. Canberra: House of Representatives Standing Committee on Regional Australia.

Bailey, C. & Pomeroy, C. 1996. Resource dependency and development options in coastal Southeast Asia. *Society and Natural Resources,* 9, 191-199.

Berkes, F. & Folke, C. 1998. Linking social and ecological systems for resilience and sustainability. *In:* Berkes, F. & Folke, C. (eds.) *Linked social and ecological systems: Management practices and social mechanisms for building resilience.* Cambridge: Cambridge University Press.

Blackmore, D. Water, salinity and the politics of mutual obligation. ABC Alfred Deakin Lecture Series, 2001 Melbourne International Arts festival, May 2001.

Connor, J., Franklin, B., Loch, A., Kirby, M. & Wheeler, S. 2013. Trading water to improve environmental flow outcomes. *Water Resources Research,* 49**,** 4265-4276

Crean, K. 1999. Centralised and community-based fisheries management strategies: case studies from two fisheries dependent archipelagos. *Marine Policy,* 23**,** 243-257.

Cummins, T. & Watson, A. 2012. A hundred year policy experiment: The Murray-Darling Basin in Australia. *In:* Quiggin, J., Mallawaarachchi, T. & Chambers, S. (eds.) *Water Policy Reform: Lessons in Sustainability from the Murray-Darling Basin.* Cheltenham, UK and Northampton, USA: Edward Elgar.

Daff 2011. Modelling the economic effects of the Murray-Darling Basin Plan: Report to client prepared for the Murray-Darling Basin Authority. Canberra: Australian Bureau of Agricultural and Resource Economics and Sciences.

Davidson, B. 1969. *Australia wet or dry?: The physical and economic limits to expansion of irrigation,* Carlton, Victoria, Melbourne University Press.

Dixon, P., Rimmer, M. & Wittwer, G. 2011. Saving the southern Murray-Darling Basin: The economic effects of a buyback of irrigation water. *Economic Record,* 87**,** 153-168.

DSEWPC. 2013. *Progress of Water Recovery Against Proposed 2750 GL SDL Reduction* [Online]. Canberra: Department of Sustainability, Water, Environment, Population and Communities. Available: http://www.environment.gov.au/water/policy-programs/ entitlement-purchasing/progress.html [Accessed 29 November 2013].

Folke, C. 2006. Resilience: The emergence of a perspective for social–ecological systems analyses. *Global Environmental Change,* 16**,** 253-267.

Griffin, R. 2006. *Water resource economics: the analysis of scarcity, policy and projects,* Cambridge, Massachusetts, The MIT Press.

Gunderson, L. H. & Holling, C. S. 2002. *Panarchy: understanding transformations in systems of humans and nature,* Washington DC.

Jordan, A., Wurzel, R. & Zito, A. 2005. The rise of 'new' policy instruments in comparative perspective: Has governance eclipsed government? *Political Studies,* 53**,** 477-496.

Kirby, M., Qureshi, M., Mainuddin, M. & Dyack, B. 2006. Catchment behaviour and counter-cyclical water trade: An integrated model. *Natural Resource Modeling,* 19, 483-510.

Krannich, R. S. & Zollinger, B. 1997. Pursuing rural community development in resource-dependent areas: obstacles and opportunities. *Research in community Sociology,* 7, 201-222.

Leblanc, M., Tweed, S., Van Dijk, A. & Timbal, B. 2012. A review of historic and future hydrological changes in the Murray-Darling Basin. *Global and Planetary Change,* 80–81, 226-246.

Loch, A., Adamson, D. & Mallawaarachchi, T. accepted. Role of hydrology and economics in water management policy under increasing uncertainty. *Journal of Hydrology.*

Loch, A., Bjornlund, H. & Mciver, R. 2011. Achieving targeted environmental flows: An evaluation of alternative allocation and trading models under scarce supply – lessons from the Australian reform process. *Environment & Planning C: Government & Policy,* 29, 745-760.

Loch, A., Bjornlund, H., Wheeler, S. & Connor, J. 2012. Allocation trade in Australia: a qualitative understanding of irrigator motives and behaviour. *Australian Journal of Agricultural and Resource Economics,* 56, 42-60.

Loch, A., Wheeler, S., Bjornlund, H., Beecham, S., Edwards, J., Zuo, A. & Shanahan, M. 2013. The role of water markets in climate change adaptation. Griffith University, Gold Coast: National Climate Change Adaptation Research Facility (NCCARF).

Loch, A., Wheeler, S., Boxall, P., Hatton-Macdonald, D., Adamowicz, W. & Bjornlund, H. 2014. Irrigator preferences for water recovery budget expenditure in the Murray-Darling Basin. *Land Use Policy,* 36, 396–404.

Mallawaarachchi, T. & Harris, M. in press. Agricultural adaptation: observations and insights. *In:* Quiggin, J., Adamson, D. & Quiggin, D. (eds.).

Mallawaarachchi, T., Mcclintock, A., Adamson, D. & Quiggin, J. 2010. Investment as an adaption response to water scarcity. *Water and the Murray-Darling Basin: Have we finally got it right workshop.* University of Queensland, Brisbane 21-22 October: Risk & Sustainable Mangement Group.

Marsden Jacob Associates, Rmcg, Ebc Consultants, Dbm Consultants, Mcleod, G. & Cummins, T. 2010. Economic and social profiles and impact assessments in the Murray-Darling Basin: Synthesis report. Canberra: Murray-Darling Basin Authority.

Marshall, N. A., Fenton, D. M., Marshall, P. A. & Sutton, S. 2007. How Resource Dependency Can Influence Social Resilience within a Primary Resource Industry. *Rural Sociology,* 72, 359-390.

MDBA 2012. Proposed Basin Plan - A revised draft. Canberra: Murray-Darling Basin Authority.

MDBC 2000. Review of the operation of the Cap: Economic & social impacts. Camberwell.

NWC 2011. Water markets in Australia: a short history. Canberra: National Water Commission.

NWC 2012. Impacts of water trading in the southern Murray-Darling Basin between 2006-07 and 2010-11. Canberra: National Water Commission.

NWC 2013. Australian water markets: trends and drivers 2007–08 to 2011–12. Canberra: National Water Commission.

Overton, I. C., Colloff, M. J., Doody, T. M., Henderson, B. & Cuddy, S. M. 2009. Ecological outcomes of flow regimes in the Murray-Darling Basin. *Water for a Healthy Country Flagship Report.* Canberra: Report prepared for the National Water Commission.

Ross, H., Driml, S. & Zarezadeh, Z. 2012. Water allocation, social change and resilience. *In:* Quiggin, J., Mallawaarachchi, T. & Chambers, S. (eds.) *Water policy reform: lessons in sustainability from the Murray-Darling Basin.* Cheltenham UK: Edward Elgar Press.

Smit, B. & Wandel, J. 2006. Adaptation, adaptive capacity and vulnerability. *Global Environmental Change,* 16, 282-292.

Tisdell, J. 2010. Impact of environmental traders on water markets: An experimental analysis. *Water Resources Research,* 46.

Walker, B., Holling, C. S., Carpenter, S. R. & Kinzig, A. 2004. Resilience, adaptability and transformability in social--ecological systems. *Ecology and Society,* 9, 5.

Wheeler, S. & Cheesman, J. 2013. Key findings from a Survey of Sellers to the Restoring the Balance Programme. *Economic Papers: a Journal of Applied Economics and Policy,* 23, 340-352.

Wheeler, S., Garrick, D., Loch, A. & Bjornlund, H. 2013. Evaluating water market products to acquire water for the environment in Australia. *Land Use Policy,* 30, 427-436.

Wheeler, S., Zuo, A., Bjornlund, H. & Lane-Miller, C. 2012. Selling the farm silver? Understanding water sales to the Australian government. *Environmental and Resource Economics,* 52, 133-154.

Wiebe, K. & Gollehon, N. 2006. Agricultural Resources and Environmental Indicators. Washington DC: US Department of Agriculture Economic Bulletin (EIB-16).

Wittwer, G. & Griffith, M. 2011. Modelling drought and recovery in the southern Murray-Darling Basin. *Australian Journal of Agricultural and Resource Economics,* 55, 342-359.

World Bank 1992. *World Development Report 1992: Development and the Environment,* New York, Oxford University Press.

In: Drinking Water and Water Management
Editor: Mason Giannino
ISBN: 978-1-63117-523-7

Chapter 9

MODELLING THE GROUNDWATER FLOW TO ASSESS THE LONG-TERM ECONOMIC COST OF IRRIGATION WATER: APPLICATION IN THE MOUDANIA BASIN, GREECE

D. Latinopoulos[1]* and I. Siarkos[2]
[1]Department of Spatial Planning and Development, School of Engineering Aristotle University of Thessaloniki, Thessaloniki, Greece;
[2]Department of Civil Engineering, School of Engineering Aristotle University of Thessaloniki, Thessaloniki, Greece

EXTENDED ABSTRACT

The continuous over-exploitation of groundwater resources due to the increased demand for irrigation water has lowered water tables in many riparian areas, threatening thus many water-stressed regions. Simulation models are useful tools for efficient water management of those areas, especially when integrating economic and hydrological information. This paper focuses on the simulation of groundwater use on a basin-wide scale, in the Moudania basin, in Greece. Namely, a two-dimensional transient flow simulation in the aquifer of Moudania was

*e-mail: dlatinop@plandevel.auth.gr.

performed, aiming first of all to assess the impact of hydraulic head distribution on the pumping depth of groundwater, as well as, its relative (long-term) effect on irrigation pumping costs. In this context, a steady-state simulation was first performed in order to get the values for certain parameters (e.g., hydraulic conductivity), as well as the initial conditions for the transient simulation. These two simulations, referring as the basic simulation, were carried out using the modular three-dimensional finite-difference groundwater flow model MODFLOW.

The economic cost of groundwater abstraction was then estimated based on the cost of energy (electricity) for pumping irrigation water. The economic analysis results showed an association of pumping costs to the depth of water abstraction, as well as to the current water consumption levels. Moreover, the range of future marginal pumping costs seems to vary spatially within the basin area, as it is directly linked to the drawdown levels. However, according to our estimates, pumping costs constitute and will continue to constitute only a small part of farmers' production costs, leading thus to inefficient use of groundwater resources and thus to depletion.

The simulation model was also used to evaluate an alternative policy scenario, with respect to the hydraulic head of the aquifer and the agricultural production of the study area (i.e., farmers' income). This alternative scenario refers to the formulation and solution of the inverse problem, by assigning hydraulic head values that are higher than a predetermined level and thereafter by determining the pumping rates of irrigation wells. The final outcome of this process would be a spatially diverse system of water permits that would minimize the resource cost, which may arise due to the depletion of the aquifer.

Keywords: Groundwater resources, groundwater modelling, pumping depth, pumping costs, spatial analysis

1. Introduction

Groundwater is the major freshwater resource of our planet. However meeting all human water needs has often resulted in an excessive and wasteful use of this resource, leading many aquifers across the earth in a state of exhaustion (Siarkos et al., 2011, Siarkos et al., 2012). The problem appears to be more intense in rural areas where the over-exploitation of groundwater for the satisfaction of irrigation needs has become more intensive. This over-exploitation, especially when combined with quality degradation, has significant ecological and economic implications. Therefore, future water management requires both environmental and economic information. The simulation of a real situation, by means of a mathematical model, contributes to this direction.

The basin of Moudania, in the regional unit of Halkidiki (Region of Central Macedonia) is a typical agricultural area of Greece where current agricultural practices are extremely dependent on groundwater resources. Water for agriculture is up to now provided free of charge and irrigation plans are almost non-existent. As a result, farmers do not attribute a significant user cost to groundwater and seem to behave myopically by overexploiting regional groundwater resources. Therefore the introduction of systems to efficiently manage irrigation water is urgently needed. This need is further highlighted by the fact that groundwater is the only source for domestic water supply in the region.

Numerical models constitute a powerful, reliable and internationally recognized tool that provides significant assistance to hydrogeological studies and management of water resources (Wang and Anderson, 1982). They serve multiple purposes, such as providing a framework for organising hydrologic data, quantifying the properties and behaviour of the systems and allowing quantitative prediction of the responses of those systems to externally applied stresses (Elango and Sivakumar, 2007). In this paper, a MODFLOW-based model was employed within the framework of the Groundwater Modelling System (GMS 8.1) to study: (a) the groundwater processes of the semi-confined aquifer system in the study area and more specifically: the distribution of hydraulic head and its impact on the pumping depth of groundwater (in the location of irrigation wells), (b) the long-term effect of pumping depth on the pumping costs of irrigation wells.

The modular three-dimensional finite-difference groundwater flow model MODFLOW (McDonald and Harbaugh, 1988) was used for the simulation of the hydraulic head (Model 1), while the economic cost was estimated based on the energy (electricity) cost of irrigation water. Moreover, an alternative scenario (Model 2) was considered and evaluated by forming and solving the inverse problem of Model 1. This scenario aims to balance water demand and supply in the study area by reducing the amount of water abstracted for irrigation purposes, resulting thus to relatively stable groundwater levels (based on this scenario, the hydraulic head remains the same throughout the whole simulation period and throughout the whole study area). Pumping costs of irrigation wells were recalculated since both hydraulic head and pumping rates of wells were modified. Irrigation water costs in both models are then compared in terms of their pumping costs, taking also into consideration income losses that can occur due to yield reductions under deficit irrigation (in Model 2).

2. DESCRIPTION OF THE STUDY AREA

The hydrological basin of Moudania extends in the South-Western part of Halkidiki peninsula (south-east of the city of Thessaloniki). It is part of a larger region known as "Kalamaria Plain", which constitutes the prime agricultural area of Halkidiki that is cultivated and irrigated intensively through the year (Figure 1). The catchment covers approximately 127 km^2, with a mean soil elevation of 211m (above the sea level) and a mean soil slope of 1.8%. It is actually a coastal basin which is in hydraulic communication with the sea, as it is bordered to the South by the Thermaikos Gulf (Latinopoulos et al., 2003, Xefteris et al., 2004, Siarkos and Latinopoulos, 2013).

In order to achieve more detailed calculations of its hydrological parameters, the basin was divided into two sub-regions, the hilly area in the North (46.3% of the basin area) and the flat area in the South (53.7% of the basin area) (Figure 1). The climate of the study area is semi-arid to humid, typically Mediterranean and the average annual precipitation is around 417 mm for the flat area and 504 mm for the hilly one. The annual recharge of groundwater accounts for 14.9% of the annual precipitation (62.3 mm) in the flat area and for 18.3% (92.3 mm) in the hilly area. Almost 80% of the region is used as agricultural land, whereas the remaining 20% is woodland. The hydrographic network is dense, especially in the hilly area, draining directly to the sea, while the most exploited part of the local groundwater resources is confined (Latinopoulos et al., 2003, Xefteris et al., 2004, Siarkos and Latinopoulos, 2013).

The whole basin is a typical rural area, where agriculture dominates both local economy and land uses and where groundwater is intensively used for irrigation. Almost 50% of the irrigated area is cultivated with olive trees. In the remaining area, the prevailing crops are: orchard trees (30% of the total area), vegetables, cotton and corn. Irrigation needs are solely satisfied by the aforementioned confined aquifer system, which is considered to be semi-confined based on the storage coefficient values resulted from various pumping tests and the fact that the soil layers in the study area are relatively permeable (Latinopoulos et al., 2003). A large number of irrigation wells (482) are densely located in the seven municipal districts of the study area as follows: Ag.Panteleimon (33), Zografou (12), Dionisiou (122), Portaria (140), Flogita (30), Simantra (94) and N. Moudania (51) (Figure 1). This distribution corresponds to the year 2001, on which all MODFLOW simulations were based (i.e., steady and transient state, as well as, the inverse problem simulation).

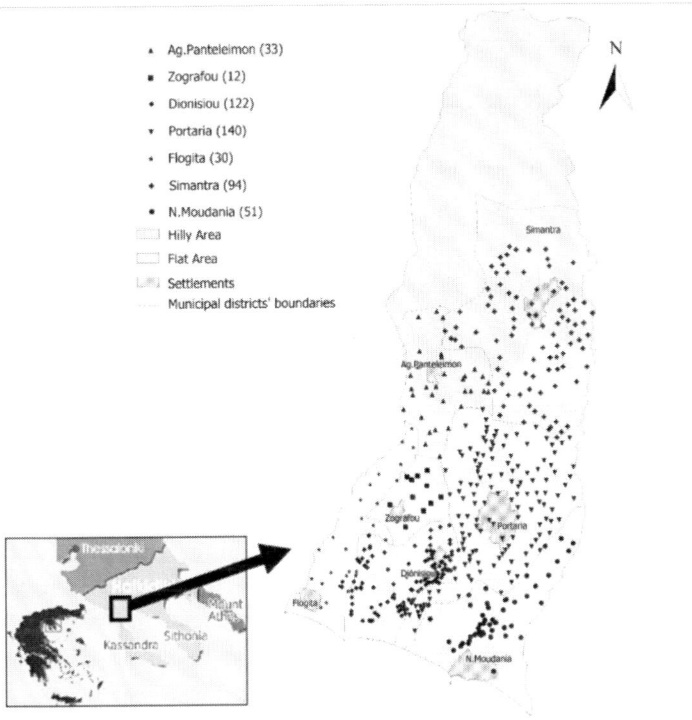

Figure 1. Location of the study area and distribution of irrigation wells.

3. Groundwater Flow Modelling

3.1. Modelling Strategies

As already mentioned, the modeling approach consists of two main steps (simulations), that is, the main simulation (Model 1) and the inverse problem (Model 2), which are described below. The main simulation model was conducted in two stages:

a) A steady-state simulation was first carried out in order to adjust specific model parameters, such as hydraulic conductivity and conductance, as well as to get the initial head values for the transient simulation.
b) A transient-state simulation was then performed to adjust the rest of the parameters (e.g., groundwater recharge, storage coefficient), as well as to observe the aquifer response (head distribution) at different time periods under different stresses (recharge and discharge). To this task, pumping/irrigation (1st May – 30th September, 153 days) and non-pumping (1st October – 30th April, 212 days) periods were considered, since both recharge and discharge receive different values in each period.

Both steady-state and transient models were calibrated using trial-and-error procedures, according to which the values of various parameters were modified until a sufficient match between the observed and the simulated water level contours was obtained.

In the case of the inverse problem, a new simulation based on the previous transient simulation was conducted. Specifically, in this model, the values of various parameters (e.g., hydraulic contuctivity, storage coefficient, recharge) except of the pumping rates of irrigation wells were kept constant, while considering at the same time a zero decline in the hydraulic head throughout the whole simulation period and throughout the whole study area1. In this way, new values of pumping rates were calculated for each group of wells belonging to the same municipal district. The model was callibrated, using the same procedure as before, but this time the observed values coincided with the initial conditions of the aquifer, since the above assumption was made.

3.2. Conceptual Model of the Study Area

Developing the conceptual model is the most important part of the modelling process. The purpose of the conceptualization is to simplify the field problem and to organize the associated field data, so that the system can be analyzed more readily. It is critical that the conceptual model is a valid representation of the vital hydrogeological conditions and involves definition

[1]This assumption was made in order to balance irrigation water demand and supply during the simulation period.

of the hydro-stratigraphic units, water balances and flow system (Kushwaha et al., 2009). The conceptualization should thus include the synthesis and framing up of data pertaining to geology, hydrogeology, hydrology, meteorology and groundwater stresses (Ahmed and Umar, 2009). In this study, the conceptual model and input data of the study area were primarily developed based upon compiled information derived from Latinopoulos et al. (2003), which includes the potentiometric surface, the hydraulic properties, as well as the recharge and discharge components.

3.2.1. Aquifer Geometry and Spatial Discritization

The model domain was discretized into a single layer finite difference grid of 120 rows and 80 columns, making a total of 9,600 cells of 150x150 meters, of which 6,555 were active and 3,045 were inactive. The layer in the Z-axis represents the semi-confined aquifer, which constitutes the main exploitable aquifer of the study area. Regarding the geometry of the aquifer, the upper limit was set at the depth at which the clay coating reaches below the ground level (an average value of 50m was obtained based on well logs), while the lower limit resulted from the upper by subtracting the theoretical thickness of the aquifer (an average value of 250m was obtained based on well logs and geological sections). This value of aquifer thickness corresponds to various successive permeable layers.

3.2.2. Boundary Conditions

The boundary conditions (Figure 2) were specified based on previous piezometric information, geological data and wells in the study area. The eastern and western boundaries are no-flow boundaries, as there is no hydraulic connection between the aquifer body and the eastern and western regions. The south and north boundaries were both simulated as general head boundaries (GHB), illustrating the lateral groundwater inflows or outflows occurring from them. The south boundary coincides with the sea level (0m), while the north boundary does not coincide with the physical boundary of the aquifer, but, based on observed data (isopiezometric map of reference year 2001, resulted by applying the Ordinary Kriging technique in groundwater levels, using data from numerous monitoring wells in the aquifer) it was set at the isopiozometric contour line of 150m, varying in time, as far as the transient simulation is concerned. This variation was based on measured data from monitoring wells located close to this boundary, as well as to data concerning the annual decline of hydraulic head.

The assumption regarding the north boundary was based on the fact that above the predefined contour of 150m there is no available and reliable data concerning the various parameters of local aquifers. What is worth-mentioning is the fact that regarding the inverse problem, the north boundary was kept constant with time, since there is no decline in the hydraulic head. Moreover, regarding this type of boundary (GHB), conductance has to be defined, thus this parameter was subjected to calibration in the steady-state simulation.

3.2.3. Aquifer Parameters

With regard to the various aquifer parameters, such as hydraulic conductivity and storage coefficient, its values were resulted from six pumping tests and therefore they were assigned to six distinct zones using the Thiessen Polygon method (Figure 3). This method is preferred over contouring where data points are sparse (Ahmed and Umar, 2009). So, the hydraulic conductivity values assigned to the model ranged between 0.125 and 0.240 m/d, while storage coefficient ranged between $3.0x10^{-2}$ and $6.25x10^{-2}$. Due to the high uncertainty of hydraulic conductivity and storage coefficient, they were subjected to calibration using the above values as initial values for the steady state and the transient simulations, respectively.

3.2.4. Recharge

Rainfall, irrigation return flows and wastewater return flows are considered to be the exclusive factors that contribute to the recharge of the aquifer. Concerning the aquifer recharge from rainfall, the Thornthwaite and Mather method (1955) was used according to which,the amount of water infiltrating into the ground is equal to precipitation minus actual evapotranspiration and direct run-off. Distinct values were determined for each sub-region and introduced to the model (0.278 mm/d for the hilly area and 0.143 mm/d for the flat area). With regard to the other two parameters (i.e., irrigation and wastewater return flows), it was assumed that 15% of irrigation water and 60% of produced wastewater (which equal to 80% of domestic water-supply) return into the aquifer. It should be mentioned that irrigation and wastewater return flows were estimated per municipal district, since significant spatial differences were found in agriculture (due to different cropping patterns per district) and domestic (due to the different population size per district) water use. It should be also mentioned that in the north-west side of the aquifer (see Figure 4) there is a relatively small area where there is neither agricultural activity nor settlements so; it is assumed that there is no irrigation and wastewater production (Figure 4). With regard to the steady-state simulation, recharge was considered constant throughout the year, while regarding transient simulation, recharge received different values depending on which period (pumping or non-pumping period) it was related to.

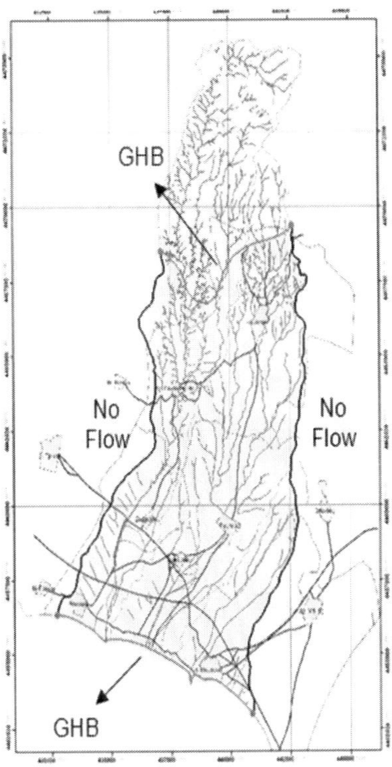

Figure 2. Model boundary conditions.

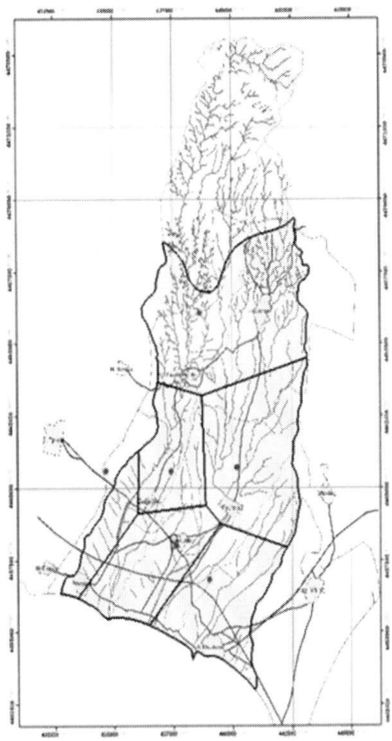

Figure 3. The six distinct zones of the study area.

3.2.5. Discharge - Water Consumption

The main discharge from the aquifer occurs from withdrawals for domestic and irrigation purposes (there is also a small percentage of water used in livestock, but it was considered as negligible and therefore it was ignored). Domestic consumption rates are estimated according to the regional average daily water consumption per inhabitant. These values are multiplied by the population of each municipality district, based on demographic data encountered in the census of 2011, taking also into account the summer population, (residents, visitors and tourists)[2]. The annual consumption of domestic water was then divided by the number of operating wells in each district in order to determine the average pumping rate per well (Table 1). The estimation of water demand for irrigation in each district was based on the current local cropping patterns and the optimum level of water applied per crop during the whole plantation period. These values are once again distributed to the individual pumping wells used for irrigation (Table 2). Figure 5 illustrates the distribution of both irrigation and domestic water-supply wells operating per district.

However, it should be mentioned that regarding the transient simulation, irrigation wells – contrary to domestic supply ones – operate only during the pumping/irrigation period, so that their pumping rate equals to zero during the rest of the year. Moreover, as presented in the following section, in the case of the inverse problem, pumping rates were further modified using a trial-and-error procedure in order to achieve the predefined environmental goals (zero drawdown).

Table 1. Estimation of domestic well pumping rates for the various municipality districts

Municipality district	Number of operating wells	Total annual consumption ($x10^3$ m^3/year)	Average pumping rate per well (m^3/d)
Ag. Panteleimon	3 wells	20	18
Zografou	2 wells	20	27
Dionisiou	5 wells	180	99
Portaria	5 wells	75	41
Flogita	7 wells	200	78
Simantra	6 wells	65	30
N. Moudania	9 wells	370	113

[2] The study area, as well as the broader Halkidiki region, attracts many tourists during the summer period. So, summer population is considerably higher when permanent residents are supplemented by seasonal residents.

Table 2.Estimation of irrigation well pumping rates for the various municipality districts

Municipality district	Number of operating wells	Total annual consumption (x10^3 m^3/year)	Average pumping rate per well (m^3/d) (steady-state)	Average pumping rate per well (m^3/d) (transient-state)
Ag. Panteleimon	33 wells	870	72	172
Zografou	12 wells	765	175	417
Dionisiou	122 wells	1,664	37	89
Portaria	140 wells	5,709	112	267
Flogita	30 wells	1,314	120	286
Municipality district	Number of operating wells	Total annual consumption (x10^3 m^3/year)	Average pumping rate per well (m^3/d) (steady-state)	Average pumping rate per well (m^3/d) (transient-state)
Simantra	94 wells	3,409	99	237
N. Moudania	51 wells	2,348	126	301

3.3. Model Calibration

The purpose of model calibration is to establish that the model can reproduce adequately field measured heads and flows (Ahmed and Umar, 2009). Trial-and-error process was used for the calibration of the models developed herein. Namely, with regard to the main simulation model (Model 1), a two-stage calibration procedure was followed. First, a quasi-steady state calibration was performed in order to adjust the initial estimates of specific model parameters - such as hydraulic conductivity and conductance – aiming to obtain the best match between simulated and actual (measured) water levels. The calibration was made using 17 observation wells monitored during September 2001. The hydraulic conductivity values ranged between 0.09 and 0.60 m/d (maximum values occurred in coastal zones), while conductance received a value of 1.1 m^2/d in the north boundary and 1.0 m^2/d in the south boundary.

The computed water level accuracy was tested by comparing the calculated mean error (ME), mean absolute (MAE) and root mean square error

(RMSE). The ME is found equal to 0.276 indicating that, on average, the simulated groundwater levels were slightly higher than the corresponding observed groundwater levels. The MAE and RMSE estimates were equal to 1.116 and 1.477 respectively.

Then, a transient state calibration was carried out during the one-year period: 1/10/2001 – 30/9/2002. The one-year period was further divided into two stress periods, each representing the pumping and non-pumping period. As before, the transient calibration was performed by using a trial-and-error process of adjusting the initial parameters (estimates) of recharge and specific storage, until a satisfactory convergence between the model results and the observed field data is obtained.

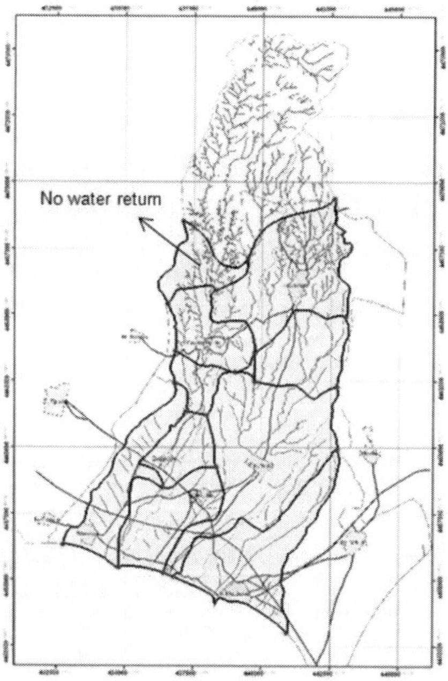

Figure 4. Various recharge zones.

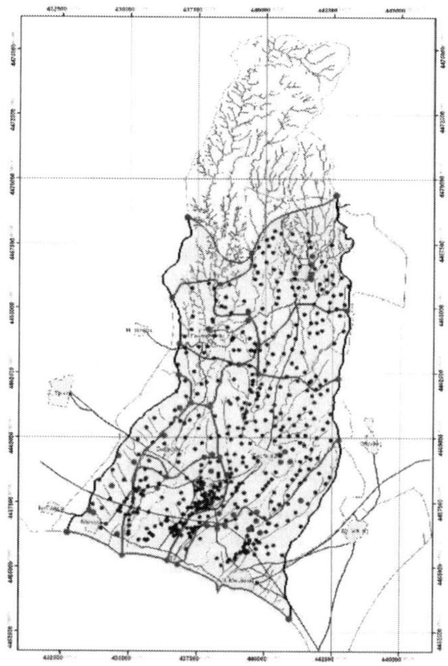

Figure 5. Location of irrigation and domestic wells.

The calibration was made using 13 (Figure 7a) observation wells, which were monitored during September 2002. The recharge values ranged between 0.216 and 0.359 mm/d in the pumping period and between 0.146 and 0.281 mm/d in the non-pumping period, while the storage coefficient values ranged between 3.75×10^{-2} and 4.5×10^{-2}. Finally, the ME, MAE and RMSE estimates were found equal to 0.615, 1.994 and 2.345, respectively.

Figure 6a shows the results of the transient model calibration in a scattergram of observed versus simulated groundwater levels, while Figure 7a shows the results in a form of a map with the calibration targets drawn next to the monitoring wells displaying the error in each well.

The largest residuals at certain wells of the study area (especially in the case of transient state simulation) may be due to anisotropy and heterogeneity of the aquifer, that is to properties (i.e., hydraulic conductivity, storage coefficient), which were not represented properly when dividing the study area into six distinct zones.

Moreover, in our analysis the effect of seawater intrusion to density variations was not taken into consideration. This is likely to be the main reason for the increased discrepancies between modelled and observed water levels in the coastal part of the aquifer. However, a perfect fit was not the goal of this study, since it primarily focuses on the generic process of the aquifer system.

With regard to the inverse problem simulation (Model 2), the calibration was performed for the same time period as in the transient simulation (1/10/2001 - 30/9/2002). Since the calculated values of the hydraulic head in this model are not the actual ones, several (virtual) observations wells were placed in areas where data is sparse (i.e., in the central and south part of the study area), as noted in Figure 7b.

Hence, the calibration was performed using 21 observation wells, which were supposed to be monitored during September 2002. Table 3 presents the modified values of average pumping rates for each group of wells, as well as the required reduction rates (%) in order to achieve the district-specific environmental targets.

As it can be noticed, the greatest water reductions (17%) occur in the municipality districts of Portaria and Zografou, where the current use of irrigation water is more intense. Finally, the ME, MAE and RMSE estimates of this simulation were found equal to -0.05, 0.423 and 0.563, respectively, indicating thus an excellent model calibration, a fact that can be seen both in Figures 6b and 7b.

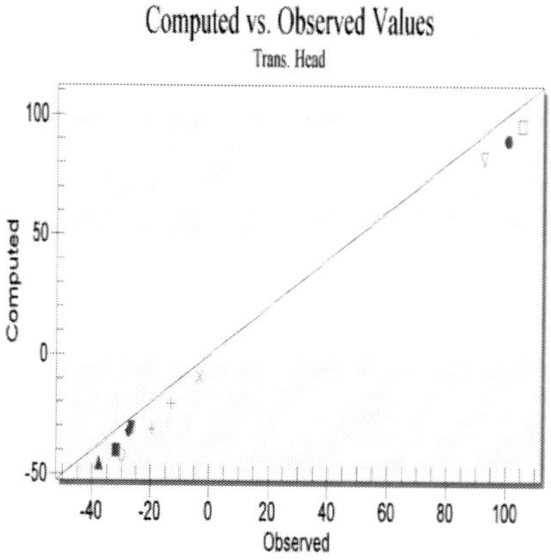

Figure 6a. Observed versus simulated groundwater levels for the transient simulation.

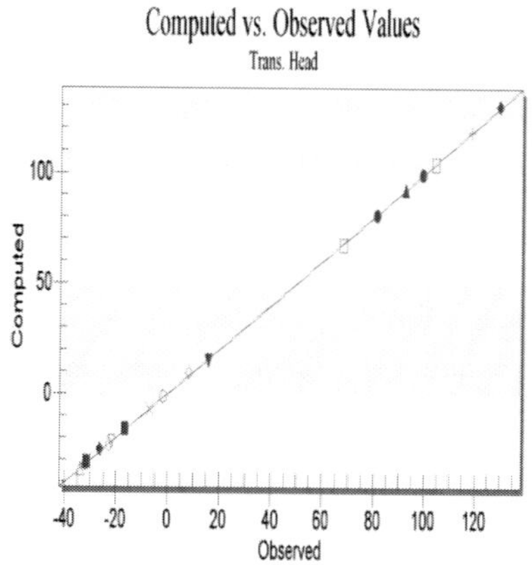

Figure 6b. Observed versus simulated groundwater levels for the inverse problem simulation.

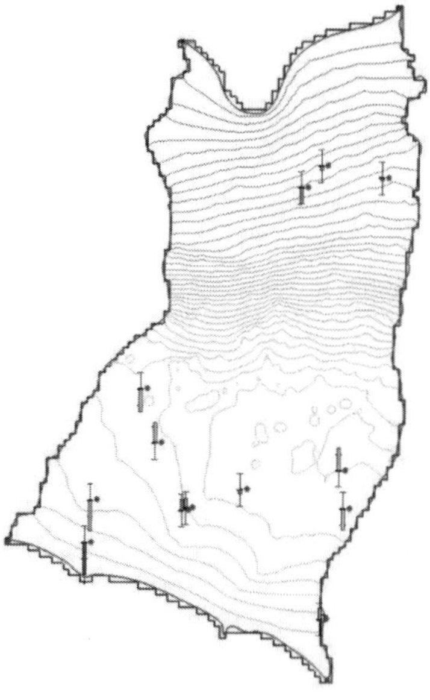

Figure 7a. Map of calibration targets next to monitoring wells (Model 1).

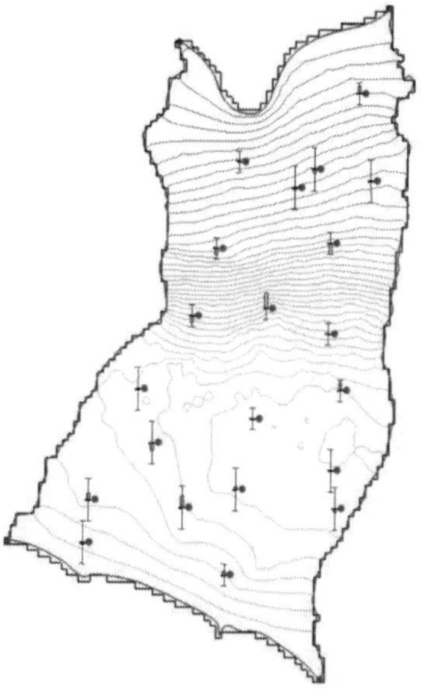

Figure 7b. Map of calibration targets next to monitoring wells (Model 2).

Table 3.Estimation of irrigation well pumping rates for the various municipality districts

Municipality district	Number of operating wells	Average pumping rate per well (m^3/d) (transient-state)	Average pumping rate per well (m^3/d) (inverse problem)	Reduction of average pumping rate (%)
Ag. Panteleimon	33 wells	172	155	10%
Zografou	12 wells	417	346	17%
Dionisiou	122 wells	89	80	10%
Portaria	140 wells	267	221	17%
Flogita	30 wells	286	243	15%
Simantra	94 wells	237	201	15%
N. Moudania	51 wells	301	259	14%

3.4. Model Results

After the calibration process, the two models were run for a 20-year simulation period (2002 - 2022) in order to observe the distribution of hydraulic head and then to estimate the pumping depth in the location of irrigation wells. Figures 8a and 8b illustrate the isopiezometric contours at the end of the pumping period, in year 2022 (which coincides with the end of the simulation period) for both simulations. The water table and pumping depth levels in Model 2 actually serve as a close representation of the current state. As expected, a gradual decline of groundwater levels, especially in areas with a high density of wells (e.g., central part of the flat area) is observed in Model 1 (main simulation model). Figures 9a and 9b show the pumping depth contours at the end of pumping period, in year 2022, for both simulations, which were estimated by subtracting the hydraulic head (Figures 8a and 8b) from the surface elevations. According to these figures, pumping depths are closely related to the drawdown (e.g., central part of the flat area), as well as to surface elevations (e.g., north part of the hilly area). Tables 4 and 5 present the forecasted pumping depths per municipal district in year 2022 for the basic simulation and the inverse problem, respectively. Comparing the average (or median) pumping depth values, it is clear that, in both simulations, Portaria and Zografou districts, which are both centrally located in the study area, are characterized by increased pumping depths due to the low hydraulic head. On the other hand, high values were also found in the district of Simantra, mainly due to its higher average surface elevation.

Figure 10 depicts the hydraulic head variation of a typical irrigation well located in the most affected area throughout the whole simulation period for both simulations. In this figure, it is worth-noting the remarkable temporal variation of hydraulic heads during each year in both models, which is indicative: (a) of the rate of aquifer's natural drainage and recharge, as well as (b) of the rate of groundwater exploitation in order to meet the regional water needs. Concerning the main simulation model, Figure 10a depicts a rapid decline of hydraulic head, reflecting the continuous depletion of groundwater reserves under current farming practices. Regarding the inverse problem (Figure 10b), this decline is minimal, as a result of reducing the overall (annual) future water consumption in agriculture with respect to the current consumption volumes. The following paragraph will focus on the economic implications of both models, trying to examine the financial cost of preventing drawdown effects in the study area.

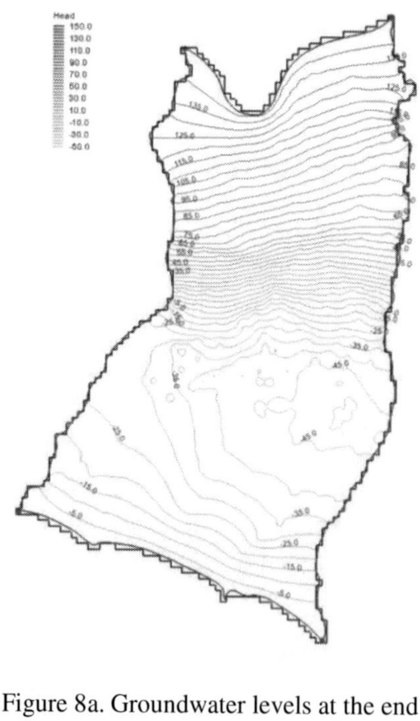

Figure 8a. Groundwater levels at the end of simulation period (2022) for Model 1.

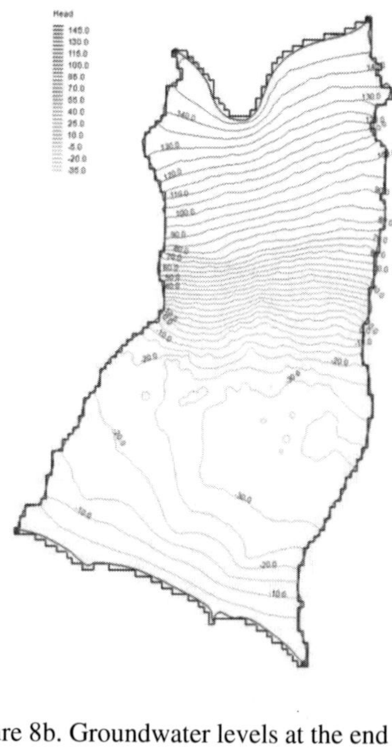

Figure 8b. Groundwater levels at the end of simulation period (2022) for Model 2.

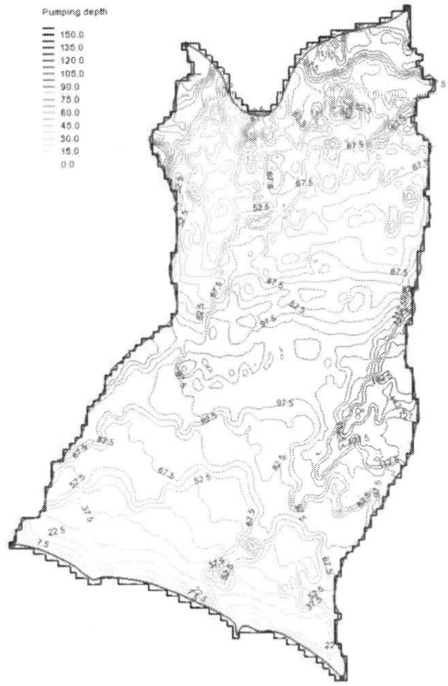

Figure 9a. Pumping depth at the end of the simulation period (2022) for Model 1.

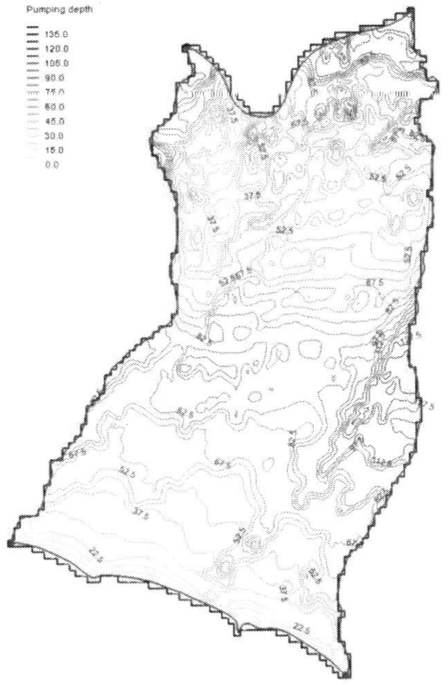

Figure 9b. Pumping depth at the end of the simulation period (2022) for Model 2.

Table 4. Pumping depth per municipal district (2022) for the main simulation (Model 1)

MunicipalDistrict	Pumping depth (m)			
	Min	Max	Average	Median
Simantra	33.10	124.95	72.92	71.24
Ag. Panteleimon	27.71	102.20	58.02	57.73
Portaria	34.51	155.92	96.24	100.50
Zografou	73.64	118.67	103.49	104.97
Flogita	21.99	108.76	61.50	57.84
N. Moudania	28.35	130.01	62.57	53.79
Dionisiou	25.71	82.85	52.54	55.24

Table 5. Pumping depth per municipal district (2022) for the inverse problem (Model 2)

Municipal District	Pumping depth (m)			
	Min	Max	Average	Median
Simantra	24.41	117.93	63.01	61.23
Ag. Panteleimon	17.50	90.17	47.59	47.14
Portaria	32.79	139.45	84.73	85.88
Zografou	65.97	106.11	92.13	92.84
Flogita	21.12	100.04	56.21	54.51
N. Moudania	27.28	118.09	57.67	49.59
Dionisiou	24.36	75.18	48.07	49.85

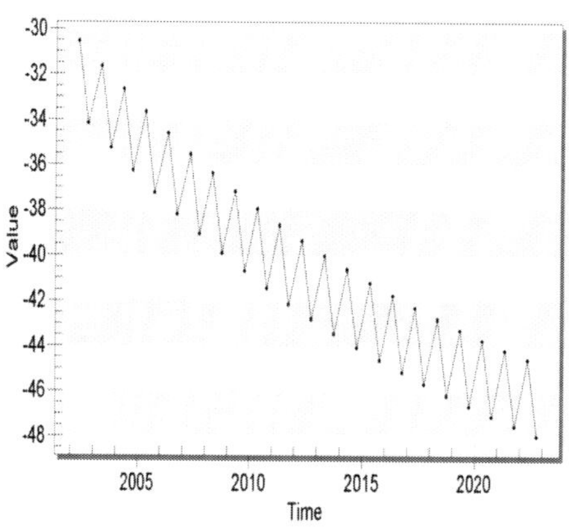

Figure 10a. Hydraulic head variation in a typical irrigation well for Model 1.

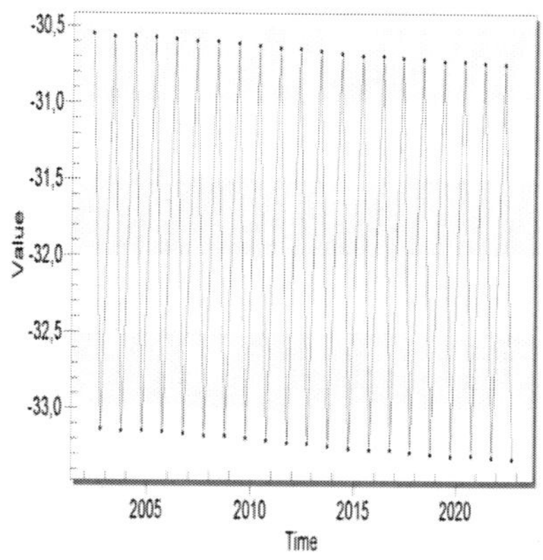

Figure 10b. Hydraulic head variation in a typical irrigation well for Model 2.

4. Economic Cost of Groundwater Management

The results of the above simulation models can be further used in order to estimate the economic costs of irrigation water management. The main simulation model (Model 1) represents a policy scenario (Scenario 1) where farmers continue to act myopically over the entire 20-year period causing an over-exploitation of the aquifer. In this scenario, farmers are only facing higher pumping costs due to the drawdown of groundwater. However, there is a significant indirect cost associated with groundwater extraction, which is often termed as resource cost (EC, 2003)[3]. On the other hand, the simulation of the inverse problem (Model 2) represents a policy scenario (Scenario 2) where local water agencies are implementing a spatially diverse system of water permits that aims to balance water demand and supply in the Moudania Basin. This scenario is associated with significant implementation costs as it presupposes the reduction of groundwater abstractions, and consequently the application of deficit irrigation strategies. A number of case-specific yield response functions can be used to assess these costs in each (municipal) district.

4.1. Pumping Costs

First of all, it is essential to estimate the pumping costs of irrigation water in both policy scenarios and then to compare these estimates in order to determine the order of magnitude of the drawdown (cost) effect through time (for the 20-year simulation period) and space. As already mentioned, the pumping costs are estimated according to the energy (electricity) cost associated with irrigation water abstractions. Namely, the following simple formula was used to calculate annual pumping costs (C) in each well:

$$C = p_e P \Delta t \tag{1}$$

where p_e=energy price (€/J), P=pumping power (Watt) and Δt=irrigation period. Pumping power per well is then estimated as follows:

$$P = d \cdot g(z_0 - h) \cdot Q \tag{2}$$

where d=water density(kg/m^3), g=gravitational acceleration (m/s^2), z_0=surface elevation (m), h=hydraulic head (m) and Q=pumping rate, which is constant in time (m^3/s).

Using these formulas, the current pumping costs per unit of water (€/m^3) were estimated in each well (Figure 11) for both scenarios. The reasoning for this is to examine the spatial variability of pumping costs throughout the region and to investigate its relation with surface elevation and hydraulic head.

[3] Resource costs are defined as the costs of foregone opportunities in time and/or space which other uses suffer due to the depletion of the resource beyond its natural rate of recharge or recovery (e.g. it is linked to the over-abstraction of groundwater).

Furthermore, the comparison between Scenarios 1 and 2 can be interpreted as the effect of local/regional drawdown to the future pumping costs.

According to Figure 11a, the average pumping costs under Scenario 1 will be higher in the municipal districts of Zografou (0.0157 €/m^3) and Portaria (0.0146 €/m^3) due to the combined effect of lower pumping levels and higher irrigation water needs (i.e., higher pumping rates). The lower values were found in the districts of Ag. Panteleimon and Dionisiou (0.0072€/m^3 and 0.0073€/m^3 respectively). On the other hand, under the inverse problem simulation (Scenario 2 – Figure 11b), pumping costs are found lower in all regions of the study area. Greater (%) cost reductions might be obtained in the municipal districts of Ag. Panteleimon and Dionisiou (-18.0% and -13.4% respectively) mainly due to the percentage changes in drawdown (see Tables 4 and 5).

In general, the myopic farmer behaviour seems to result to higher future irrigation costs in an order of magnitude that is almost negligible (i.e., from 8% up to 18% of the current pumping costs, which constitute only a small part of farmers' production costs, ranging from 2.5 up to 13.5 € per hectare). Within this framework, future individual farming (irrigation) decisions will inevitably lead to wasteful use and depletion of groundwater resources. It should be noted that the current analysis assumes that the constant decline of hydraulic head will not significantly alter salinity conditions (due to seawater intrusion) during the simulation period. Otherwise, the effect of water salinity to yield production and hence to farmers' income may be substantial, particularly in wells close to the seashore.

4.2. Water Permits

The second policy scenario assumes that water agency/agencies will act rationally by assigning a quantity of non tradable water permits (quotas) to local farmers seeking to maintain a balance between sustainable abstraction and recharge. These permits are supposed to be issued annually, in proportion to the historical rights of irrigators in each (municipal) district. In particular, the proportion of water reductions would be equal to the reduction rates (%) from Table 3, which can achieve the district-specific targets imposed by Model 2. This process ensures equity of water resource allocation in each region.

Given the high percentage of permanent crops, that is, crops that are actually associated with high investment costs, the restructuring of the area's cropping plans seems to be an economically inefficient solution. For this reason, it is likely that farmers are going to address water availability constraints by means of deficit irrigation strategies. The computer software CROPWAT (Smith, 1992) can estimate water requirements, irrigation schedules for deficit irrigation and yield losses for various deficit scenarios (as a % of maximum yield). For this, it uses the yield response factors derived from the crop-water production functions synthesized in FAO (Doorenbos and Kassam, 1979):

$$\frac{Y_a}{Y_m} = \prod_{i=1}^{N} \left(1 - ky_i \left(1 - \frac{ET_{a_i}}{ET_{c_i}} \right) \right) \tag{3}$$

where Y_a is the actual harvested yield, Y_m is the maximum crop yield (under given management conditions that can be obtained when water is not limiting), N is the number of growth stages, ETa_i is the actual evapotranspiration and ETc_i is the maximum evapotranspiration for non-limiting water conditions during the i-th stage of growth, and finally, ky_i is the yield response factor of crop during the i-th stage.

This paper follows the procedure described in Latinopoulos (2005) in order to derive the yield responses for the main crops in the study area. Namely, the actual water requirements for each crop were estimated using the CROPWAT software, according to local climate, soil and crop characteristics, as well as, to water application efficiency and irrigation scheduling. It should be noted that the maximum crop yield values (Y_m) were found from secondary sources (local agricultural agencies). Then the pre-determined water reduction levels in each municipal district were simulated. These water reductions may either occur continuously over the total growing period or may occur during any one of the individual growth periods. For the benefit of computation easiness it was assumed that they were equally apportioned over the whole period. Under this assumption, the values of Y_a were estimated for the selected levels of water reductions. Once the actual yield under deficit irrigation is estimated it is possible to calculate the associated (reduced) gross margin for each crop (farmer's income).

The difference of gross margin between Scenario 1 (without water constraints) and Scenario 2 (with water constraints/permits) indicates the income losses due to yield reductions under deficit irrigation. However, in order to find the precise cost of water management in irrigated agriculture, it is essential to subtract from these income losses the avoided pumping costs

(Scenario 2, as already mentioned, always results to lower pumping costs than Scenario 1). Figure 12 shows the aggregate income losses and avoided pumping costs per municipal district, as well as the average per hectare implementation cost of Scenario 2.

According to these results, it is apparent that pumping costs were able to make little impact on irrigation water management decision-making. Figure 12 also demonstrates that the implementation of a quota system will be more expensive in absolute terms (€/district) in the eastern part of the Moudania Basin, as a result of: (a) higher pumping rate reductions, (b) largest irrigated area and (c) more sensitive crops to water deficits. Furthermore, it is obvious that the per hectare implementation costs exhibit a significant spatial variability, ranging from 25€/ha in Ag. Panteleimon up to 190€/ha in Portaria.

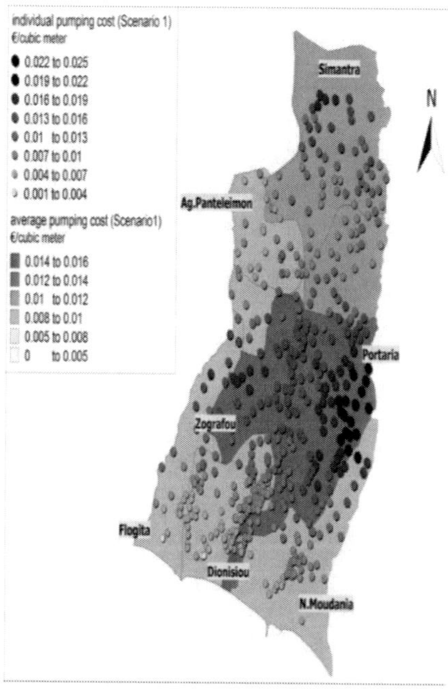

Figure 11a. Spatial variability of irrigation pumping costs (Scenario 1)

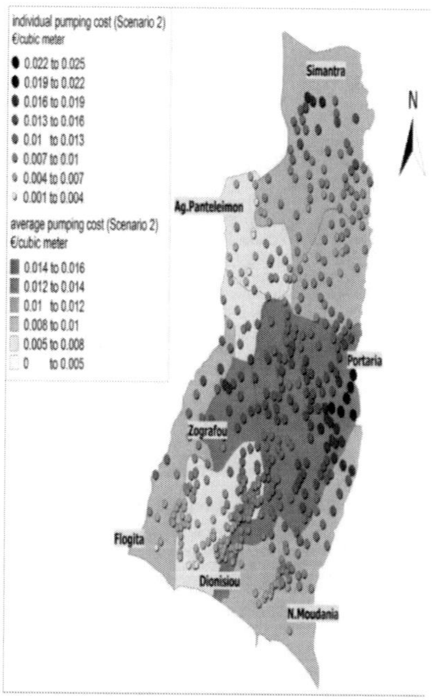

Figure 11b. Spatial variability of irrigation pumping costs (Scenario 2)

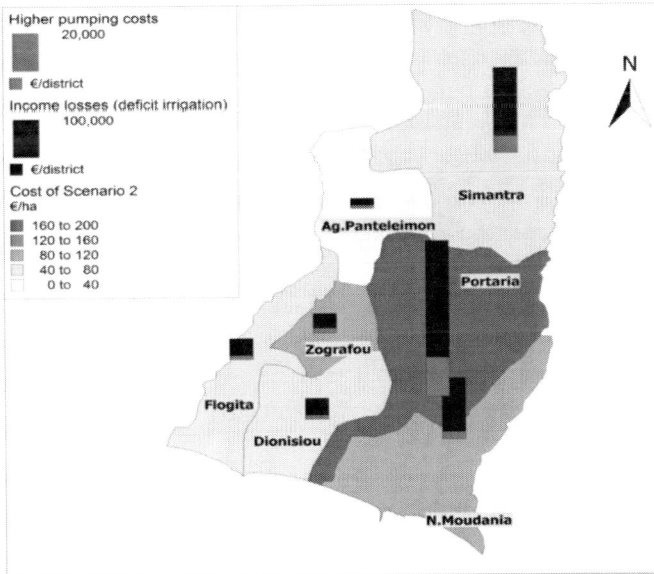

Figure 12. Implementation cost of Scenario 2.

This outcome indicates that water agencies should try to find a mechanism that will safeguard not only the intra-district equity but also the inter-district equity. A possible solution for this is to use a tradable permit system, where farmers will be able to trade their quotas with other farmers inside the Basin.

CONCLUSION

In this study both steady and transient models were developed, calibrated, and used to simulate groundwater flow in Moudania aquifer. The analysis aims to estimate the hydraulic head values required for the determination of pumping depth and subsequently of pumping cost per well.

Regarding the groundwater simulation referring as basic simulation, increased drawdown is observed especially in the central part of the region. If current irrigation usage and practices continue, a considerable water-level decline will take place in the area, causing the depletion of groundwater reserves.

Spatial economic analysis results showed - as expected – an association of pumping costs to: (a) the depth of water abstraction, as well as (b) to the current water consumption levels. Moreover, the future variation in pumping costs seems to reflect the regional drawdown contours.

The individual farmer behaviour seems to result to higher future irrigation costs in an order of magnitude that is almost negligible with respect to farmers' cultivation costs (i.e., from 8% up to 18% of the current pumping costs). Within this framework, future farming decisions will inevitably lead to wasteful use and depletion of groundwater resources. On this account, an alternative policy scenario was taken into consideration, according to which no temporal and spatial decline in the hydraulic head (i.e., at the pumping depth) is permitted. The evaluation of this scenario was carried out by solving the inverse problem of the transient simulation.

The results showed that the recommended reduction of pumping rates varies among regions in a range between 10% and 17%, the largest of which occurs where water abstraction is currently more intense. On the other hand, the economic implications of adopting a water constraint strategy (using a quota system), seem to be quite expensive and unequal distributed among the various regions. However, these costs should be further examined with respect to the "resource costs" - that is, to the current and future opportunity costs of water supplies - in order to make future decisions about water resources conservation.

For future research, more thorough study of some aquifer (e.g., thickness, boundary conditions, hydraulic conductivity, storage coefficient, and recharge) and agricultural parameters (e.g., crop pattern and irrigation method at the farm level) should be done to achieve more adequate calibration regarding the transient model.

A number of alternative policy scenarios could be also examined and evaluated, by relaxing the zero drawdown restriction. Furthermore, an inter-district water market model could be applied to confront inter-district inequality and to maximize water use efficiency throughout the study area.

ACKNOWLEDGMENT

Part of this work was presented at the 13th International Conference on Environmental Science and Technology (CEST, 2013), held in Athens, Greece (5-7 September 2013).

REFERENCES

Ahmed I. and Umar R., 2009.Groundwater flow modelling of Yamuna–Krishniinterstream, a part of central Ganga Plain Uttar Pradesh. *Journal of Earth System Science*, Vol.118, No.5, pp. 507-523.

Doorenbos, J., and Kassam A. H., 1979. Yield response to water. Rome: FAO, 193p. Irrigation and drainage paper 33.

European Commission, 2003. Economics and the environment: the implementation challenge of the Water Framework Directive, WATECO: a guidance document.

Elango L. and Sivakumar C., 2008.*Regional simulation of a groundwater flow in coastal aquifer, Tamil Nadu, India.* In: Shakeel A., Ramaswarny J. and Abdin S. (eds), Groundwater dynamics in hard rock aquifers, New Delhi: Capital Publishing Company, pp. 234-242.

Kushwaha R., Pandit M. and Goyal R., 2009. MODFLOW based groundwater resource evaluation and prediction in Mendha sub-basin, NE Rajasthan. *Journal geological society of India*, Vol.74, pp. 449-458.

McDonald M. and Harbaugh A., 1988.*A modular three-dimensional finite-difference ground-water flow model: Techniques of Water- Resources Investigations*. U.S. Geological Survey, Book 6, Chapter A1.

Latinopoulos D., 2005. Derivation of irrigation water demand functions through linear and non-linear optimisation models: application to an intensively irrigated area in northern Greece. *Water Science and Technology: Water Supply,* Vol 5, No 6, pp 75–84, IWA Publishing.

Latinopoulos P., Theodosiou N., Xefteris A., Mallios Z., Papageoriou A. and Fotopoulou E., 2003. Developing water resources management plan for water supply-irrigation. Final Research Project Report prepared for Municipality of Moudania (in Greek).

Siarkos I., Katirtzidou M. and Latinopoulos D., 2011. Establishing cost-effective wellhead protection zones to control nitrate pollution from agriculture activities. *Proceedings of the 12th International Conference on Environmental Science and Technology*, 8-10 September, Rhodes, pp. 1724-1731.

Siarkos I., Latinopoulos D. and Katirtzidou M., 2012.*Delineating cost-effective wellhead protection zones in a rural area in Greece. Water and Environment*, doi: 10.1111/wej.12012.

Siarkos I. and Latinopoulos P., 2013. Delineation of wellhead protection zones for the control of point pollution sources in the aquifer of N. Moudania. *European Water Journal, Vol.*40, pp. 3-17.

Smith, M. 1992. CROPWAT - A computer program for irrigation planning and management. *FAO Irrigation and Drainage Paper* No.46, Rome, Italy.

Thornthwaite C. W. and Mather J. R., 1955.*The water balance*. Publications in Climatology 8(I): 1-104.

Wang H. and Anderson M., 1982.*Introduction to groundwater modeling*, W.H. Freeman, New York, USA.

Xefteris A., Anastasiadis P. and Latinopoulos P., 2004.Groundwater chemical characteristics in Kalamaria plain, Halkidiki peninsula, Greece. *Fresenius Environmental Bulletin*, Vol.13, pp. 1158-1167.

INDEX

B

C

F

G

H

I

J

K

L

M

N

O

P

Q

R

S

T

U

V

W

X

Y

Z